SAM GOUDSMIT

AND THE HUNT FOR
HITLER'S ATOM BOMB

SAM GOUDSMIT

AND THE HUNT FOR HITLER'S ATOM BOMB

Martijn van Calmthout

Translated and Edited by Michiel Horn

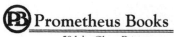

Prometheus Books

59 John Glenn Drive
Amherst, New York 14228

Published 2018 by Prometheus Books

Dutch title: *Sam Goudsmit: zijn jacht op de atoombom van Hitler*. Published in 2016 in Dutch by Meulenhoff bockerij (Amsterdam). Published by special arrangement with Meulenhoff Boekerij B. V. in conjunction with their duly appointed agent, 2 Seas Literary Agency.

The publisher gratefully acknowledges the support of the Dutch Foundation for Literature.

Nederlands
letterenfonds
dutch foundation
for literature

Cover image © AIP Emilio Segrè Visual Archives, Goudsmit Collection
Cover design by Jacqueline Nasso Cooke
Cover design © Prometheus Books

Inquiries should be addressed to
Prometheus Books
59 John Glenn Drive
Amherst, New York 14228
VOICE: 716–691–0133 • FAX: 716–691–0137
WWW.PROMETHEUSBOOKS.COM

22 21 20 19 18   5 4 3 2 1

Library of Congress Cataloging-in-Publication Data

LCCN 2018025568 (print)
ISBN 978-1-63388-450-2 (hardcover) | ISBN 978-1-63388-451-9 (ebook)

Printed in the United States of America

For Esther Goudsmit

*Those who survive never again live for themselves alone,
but always also for the dead.*

—György Konrád

*There are known knowns; there are things we know we
know. We also know there are known unknowns; that
is to say we know there are some things we do not know.
But there are also unknown unknowns—the ones we
don't know we don't know. And if one looks throughout
the history of our country and other free countries, it is
the latter category that tend to be the difficult ones.*

—Donald Rumsfeld, United States Secretary
of Defense (2001–2006)

CONTENTS

CONTENTS

10

EDITOR/TRANSLATOR'S NOTE

In translating Martijn van Calmthout's fascinating biography of Samuel Goudsmit, I incurred two debts. The first is to Martijn himself, who cheerfully responded to my frequent questions and comments. The second is to my wife, Cornelia Schuh, who is also my volunteer in-house copy editor. She read the entire text and often improved it. I am grateful to both of them.

In addition, I want to thank Lauren Humphries-Brooks at Prometheus Books, who did an excellent job copyediting the text.

Occasionally it proved impossible to obtain the original English text of passages that Martijn had translated into Dutch. In these cases I had to translate them back into English, which is, of course, not standard practice. These instances are indicated in the book with an asterisk.

Michiel Horn
Toronto, ON, May 2018

PREFACE

As was the custom in those days, in the summer of 1925 Leiden University organized a festive event in honor of Professor Hendrik Antoon Lorentz, elder statesman of theoretical physics and a national celebrity. It had been fifty years since Lorentz had obtained his doctoral degree from Leiden, and that called for a celebration. Leading scientists had come from around the world for the occasion. Among them were Albert Einstein from Berlin, father of the theory of relativity, and from Copenhagen, Niels Bohr, who in 1913 had grasped the makeup of the hydrogen atom.

Seated in the first row of the auditorium of the Kamerlingh Onnes Laboratory, the two men listened attentively to the respectful and learned speeches. But a moment came when they turned around and allowed their host, Paul Ehrenfest, to point out two students to them, Samuel Goudsmit and George Uhlenbeck. The students did not know where to look. Only when it was clear that Ehrenfest wanted to introduce his two students to Einstein and Bohr did they relax a little. Could the gentlemen elucidate their recently published article about the rotating electron a bit more closely? Of course, said Uhlenbeck, the more self-assured of the two. They can certainly do that.

During the next few days, while the festivities continued, the students spoke a few more times with the two great men about the new concept of electron spin, as the rotation was immediately called in those days. Einstein and Bohr recognized it as a major breakthrough, even if the mathematics needed to be tightened up.

The 1925 encounter between the two timid Leiden science students and the giants Einstein and Bohr, also described in Martin Klein's unfinished biography of Paul Ehrenfest, was, when I read it, the first time that the name of Samuel Goudsmit came to my attention. More than ten years ago I was working on a popular-science fictionalized biography of Albert Einstein, in which his Leiden connections intrigued me. Einstein was, so it appeared, so much at home in Leiden that he had his own room in Paul Ehrenfest's house on Witte Rozenstraat, with a backup violin for sociable entertainment during the long winter evenings, accompanied by Ehrenfest at the piano.

Every physics student knows the concept of electron spin. The rotating electron is the key to understanding atoms and the light they emit. In reality there is, of course, nothing resembling a small rotating sphere. The student learns that only quantum theory can properly describe the spin. But what is generally missing is the story of the discovery of the electron spin. Not in a laboratory or theoretical institute in a distant foreign country, but simply in Leiden, Holland. The reading room where it happened is still there. You can go visit it, and it is, in fact, incomprehensible that there is no commemorative plaque on the door. At the same time, this is characteristic of the lack of interest most physicists show in historical matters.

That explains in part why no biography of Goudsmit (1902–1978) has ever been published, even though he is one of the founders of modern physics. No matter how important electron spin is from the point of view of physics, its discoverers have remained a footnote to history.

With the passage of time this became steadily more incomprehensible. Goudsmit (with his fellow student Uhlenbeck) proved to be more than a physicist with a pioneering idea. His personal life was eventful as well as intriguing. In the 1920s, this son of a Jewish middle-class family from The Hague seized the chance to emigrate to the distant United States of America and, joined by his young wife, to make a new life in Michigan, instead of "having to become a teacher in some provincial town," as he put it. In the years that followed he took part in the headlong advance of modern physics in America. At the same time, political storm clouds gathered over Europe, and as a Jewish intellectual he decided not to return, even when the opportunity presented itself.

All the more difficult was the adventure into which the Second World War would plunge him. In 1944 and 1945, Goudsmit led a reconnaissance mission into France and Germany, both of which were badly damaged by military action, in search of a potential Nazi atom-bomb project. He had to take into custody men with whom he had collaborated scientifically in an earlier life, with Nobelist Werner Heisenberg as the most important figure. After the war, the European atom-bomb adventure turned Goudsmit into one of the leaders in the discussions about science and nuclear weapons. At the same time, he was from the 1950s the leading force in America behind a new scientific journal, *Physical Review Letters*, which would soon develop into the most important physics journal worldwide.

Sam Goudsmit did not play a leading role in the history of science, but he always played an important supporting role, as a cue-giver, prompter, conscience, and gadfly. This has had the result that he occasionally shows up in many significant scientific stories and then seems to vanish just as quickly. For that reason, I haven't picked up a history of science or biography of a scientist during the last ten years without checking the index for "Goudsmit, Samuel." Then it appears that he is present everywhere, without a high opinion of himself, and in many ways more marked by life than he was willing to admit to friends.

In 2013, the American Physical Society, an association of physical scientists in the United States, made the complete Sam Goudsmit Papers publicly available online. This archive is immense. There are hundreds of boxes with letters, documents, notes, and newspaper clippings, altogether extending over dozens of yards. They are, in fact, easier to deal with digitally than in the basement of the archives in Washington, DC. Whoever tracks through that archive systematically discovers the emotional man behind the brief appearances in the great stories of twentieth-century science. He was a talented scientist but above all talented in organizing science and in coaching a new generation of scientists. He himself did not have a high regard for his own theoretical abilities; it had, he said, actually all been no more than intuition and hand-waving.

This biography of Sam Goudsmit was ultimately made possible by the existence of the Goudsmit Papers and a decade of profiles in other historical studies. It is simultaneously intriguing and melancholy. Goudsmit was known as a jovial man, warm with people, with an agreeable manner. That upbeat exterior masked, it appears, a much more somber Sam Goudsmit whose life had been strongly marked especially by the war. Among other things, it is due to the conscientious assistance of Goudsmit's daughter, Esther Goudsmit, Professor Emerita of Biological Sciences, Oakland University, Rochester, Michigan, that I have been able to obtain more insight into that side of his life. After initial hesitation she made available to me, among other items, a family history that Sam Goudsmit had written for her in 1973. This history was composed of forty-five sheets of lined paper, in order, he wrote to his daughter, that the story of their ordinary Jewish family would not be lost.

My thanks in preparing this work goes first of all to Esther Goudsmit, daughter of Sam and Jaantje. Daniel Henderson of New York University and Daniel Kleppner of Cambridge University also offered important recollections and documents. In the course of the years, many people have helped me and shared their thoughts, some without even knowing it; among them are the historians of science Dirk van

Delft, Anne Kox, and Jeroen van Dongen. Thanks as well to Thijs Bartels, my editor at Meulenhoff, who knew how to turn an interminable project into a streamlined operation. Finally, Sam's story would never have been told without the curiosity and red pen of Mieke, the constant force in my own life.

Martijn van Calmthout
Amsterdam, 2016

A HOME STRIPPED BARE

The Hague, September 1945

I t is very early on a quiet and windy Sunday, September 23, 1945, as an American army jeep drives into the awakening downtown of The Hague. There is a threat of rain in the air; the sky above Scheveningen, west of the city, is dark. The streets are virtually empty. The Hague had been liberated in May. In many places, trees have been cut down. Here and there façades are badly damaged. Sam Goudsmit has pulled up the collar of his green army coat and has put his helmet beside him on the passenger seat. As he drives through the city, now looking so shabby, he sometimes has difficulty recognizing his place of birth. He does smell the sea in the distance; that is unchanged.

Although already several months in the past, the war is still omnipresent. Some distance to the north, behind The Hague's Central Station, well beyond Goudsmit's view, the Bezuidenhout residential area lies totally destroyed, the result of a British map-reading error in March—the bombing attack had been intended for the German V2 rocket installations in the Hague Woods. In the meantime, the largest piles of rubble have been cleared. The hundreds of dead have been buried. Farther on in Scheveningen, the first bunkers and tank obstacles of the Atlantic Wall have already been torn down, and the beach has become more or less usable by seaside visitors.

But that is in the distance. Here in the downtown area the network of streets is imprinted in Sam's brain. And so he steers the jeep almost by autopilot via the Schilderswijk area into Prinsengracht, then into Jan Hendrikstraat, along Torenstraat past the Grote Kerk, and then right, to Korte Molenstraat. So as not to attract attention, he parks the jeep on a side street and walks toward the intersection with Prinsestraat. In his memory it is a lively shopping street along which, in his early youth, a horse-drawn streetcar ran.

Even before he reaches the corner, he sees the destruction. The great, gleaming showroom windows of 86 Prinsestraat—Goudsmit-Gompers Ladies' Hats—have disappeared, and he faces the empty, gaping storefront where, before the war, his mother's fashion boutique was located. Here in 1923 he first met his wife, Jaantje, who was apprenticing at the time as a hat designer.

Only then does he see that most other properties around him are empty and stripped. They were dismantled during the long, icy "Hunger Winter" of 1944–45, when the townspeople were deprived of coal or other fuel. Everywhere the woodwork has disappeared, from the stairs to the dignified paneling. Somewhat to the left is the small, yellow house—number 82—where he was born on Friday, July 11, 1902, a chilly, gray day that hardly seemed like summer. The house is not in much better shape; even the front door is gone.

Sam Goudsmit is a forty-three-year-old Dutch-born physicist who has just made a long journey from Paris into the Netherlands. As part of the American army, he has been hunting for German scientists during the closing stages of the war. He has spoken with military security agencies and key persons at Leiden University about what is already being referred to as the "Cellastic affair." During the war, Dutch scientists and engineers, especially from Leiden, worked with the Germans in a scientific espionage network directed from Paris and concealed by a Dutch trading firm named Cellastic. Sam has determined that many academics within and outside Leiden were deeply shocked by the collaboration. Big names have been mentioned, among them that of Professor Wander de Haas, whom Sam knew well during his student years in Leiden in the 1920s.

But on this Sunday, Sam Goudsmit has summoned up all his courage to see with his own eyes what has remained of his life before the war. On the road to The Hague, he imagines for a moment how his father and mother, both looking a bit older, will be waiting for him in the store, just the way they did all those times before the war, whenever he was visiting briefly from a distant America. Or otherwise at home, at 137 Koninginnegracht, the house where he grew up, and where he lived with his sister, Rachel, four years older, and his parents. They were a Jewish middle-class family: his mother, Marianne Gompers, was the owner of the fashion boutique on Prinsestraat; his father, Isaac Goudsmit, was a successful dealer in elegant mahogany toilet seats.

As he approaches the city, he is aware that he has been entertaining an illusion, a wish-fulfillment dream. In reality he received a farewell letter from his parents early in 1943, postmarked Westerbork, written in the course of their deportation from The Hague. After that they were transported to Auschwitz on February 10, to be gassed immediately upon their arrival a day later, on his father's seventieth birthday. Goudsmit had already inferred this a couple of months earlier from Nazi documents he saw in Berlin in the summer of 1945. Ultimately the city of The Hague will officially enter their deaths in the municipal registry only in 1953, when all communications from the Red Cross have at last been processed.

Sam, upset, walks back to his jeep and starts the engine. He turns into Prinsestraat and drives at high speed along the fence of the park behind Noordeinde Palace, goes right along Hogewal and Mauritskade until he reaches the long Koninginnegracht. To the right lies Malieveld Park and behind it in the distance the devastated Bezuidenhout, where surprisingly the church tower still survives. Koninginnegracht, with its endless row of stately mansions, is almost unrecognizable because of an antitank ditch that was dug there during the war as part of the Germans' Atlantic Wall. Sam turns left and proceeds along the ditch until, a short distance before it enters Scheveningen, it follows the curve he knows so well. To the left just before that point, the street rises a bit. Sam stops in front of 137 Koninginnegracht, a house with a high porch. There his spirits sink.

The house, once a place of sunlight and stained-glass windows, of books and family evenings with aunts and uncles around the piano in the living room, has been totally stripped. There is broken glass everywhere; the frames of the two windows alongside the front door are gone, as are those of the three windows on the second floor. Still higher, on the third floor, curtains are blowing from the low windows below the eaves. Sam enters. There is a gaping hole where the wooden staircase to the second floor used to be; doors are missing; door jambs, too. Everything that was combustible had been stripped out and taken away during the previous winter. Parts of the stucco ceilings have been taken down for the combustible laths and straw. On the floor of his emptied childhood room he finds one of his old report cards, which his parents saved for all those years. "The little garden in back of the house looked sadly neglected," Sam later recalls. "Only the lilac tree was still standing." Remarkably enough, it had survived the hunt for firewood.

A wave of guilty feeling washes over Sam Goudsmit, tears in his eyes. His remorse is devastating. Perhaps, he later notes with barely concealed despair in his 1947 memoir, *Alsos*, he could, as a US citizen, have saved his parents. After all, the entry visas for them had been lying ready, everything had been prepared. But four days after the documents for Isaac and Marianne Goudsmit had been completed, the Germans had invaded the country, and it had been too late for any rescue. "If I had hurried a little more, if I had not put off one visit to the Immigration Office for one week, if I had written those necessary letters a little faster, surely I could have rescued them from the Nazis in time," he writes about the despair he felt in The Hague two years earlier.

On the other hand, he realizes, his feeling of guilt matches that of innumerable others who, just like him, lost their loved ones to the Nazis and were unable to do anything about it. "My parents were only two among the four million victims taken in filthy, jam-packed cattle trains to the concentration camps from which it was never intended they were to return."

In many ways, the return to Koninginnegracht in September 1945 is the key event not only in Goudsmit's book *Alsos*, but also in his entire life. Before the war, he was always a relatively indifferent scientific figure, who almost nonchalantly allowed life to lead him where it would. He had no very high opinion of his own scientific capacities; he believed that in his case matters depend on intuition more than on genuine insight and sound technique.

But if others asked him to do his bit, he was certainly not the most diffident. That is how, upon leaving Leiden, he landed with the Nobelist Pieter Zeeman in Amsterdam. That is how, on the invitation of his professor Paul Ehrenfest, he brought his fellow student George Uhlenbeck up to speed in physics, in passing shaking up the field in the 1920s. That is how, now suddenly famous, he left for North America in 1927, accompanied by his young wife Jeanne (Jaantje) Logher and by Uhlenbeck. (A year younger than Sam, Jaantje was born and raised in modest circumstances in Amsterdam.)

In 1943, Sam, working at MIT while on leave from his professorship in nuclear physics at the University of Michigan, is recruited by the American authorities for a reconnaissance mission in a war-torn Europe. The mission is given the name "Alsos." Operating under military command, Goudsmit and a team of fellow scientists are charged with the task of finding out, in strictest secrecy, how far the Germans have advanced toward the development of a nuclear weapon.

The question seems appropriate, certainly at that time. In 1938–39, German physicists such as Otto Hahn and Lise Meitner discovered nuclear fission. What the Germans have subsequently done with that discovery is unclear when the United States enters the war. Sam seems tailor-made for the job. Not only is he an expert on nuclear physics and does he speak fluent German, but as a graduate student he repeatedly worked in Germany, knows his way around, and is acquainted with all the kingpins in German physics. Werner Heisenberg, the leading German physicist at that point, actually stayed with the Goudsmits in Ann Arbor more than once, for the last time in the summer of 1939, shortly before war broke out in Europe. The Goudsmits' daughter, Esther Marianne, five years old, sat on his lap at the time of that last visit.

Several decades later, in 1973, Sam will write a series of letters totaling forty-five pages to his daughter, Esther, describing the family history of the Goudsmits. Esther Goudsmit, then almost forty years of age, is an assistant professor of Biological Sciences at Oakland University in Rochester, Michigan. "Dear Els," the first epistle begins in Sam's extraordinarily regular hand. He is writing on lined paper, sent, the envelope tells us, from Brookhaven National Laboratory on Long Island. "I think it would be a good idea for you to become acquainted with and part of the Goudsmit tribe. You will meet people who are very different from what you are accustomed to in your academic cocoon of biologists. Did you know that you and I are the only two Goudsmits who completed high school and, not only that, then went on to university?'"

Halfway through the letters, after long excursions about uncles and aunts, grandparents, cousins, and even unrelated Goudsmits, Sam writes about his sister, Ro (Rachel), four years older than he. How they walked from Prinsestraat together to attend Mr. Valken's school. Now and then Sam took along a fishnet to scoop up small fish in a canal that they passed. The school adjoined the drill grounds of the Royal Palace, where the cavalry went through their drills on beautiful days. "Sometimes the lessons were agreeably interrupted by excellent percussion music and jumping horses."

Home, then still at 82 Prinsestraat, is in his memory always warm and cozy. "My parents loved to have people around. On Fridays we had a sort of open house, with music. Friends and acquaintances came by and got tea and fruit and cookies, and we listened to the piano, violin, cello, trios and songs. The music was generally classical, but not very highbrow, mostly Brahms and Beethoven, songs by Schubert. Later, during the First World War, there were also popular patriotic songs. These were enjoyable evenings with lively conversation, in which Ro participated more than I did, younger and an egghead, and somehow less at ease around people."

Meanwhile, all that is now thirty years ago and farther away than ever, Goudsmit realizes as he stands in his stripped parental home. Well over a year of wandering through a Europe ravaged by war, under sometimes difficult and even ticklish circumstances, right behind and sometimes even ahead of the Allied front in Germany, turned Sam Goudsmit into a different person. The loss of his parents and his own impotence took the sparkle and uninhibited enthusiasm out of him. "In time Sam recovered from the horror, but he never again became his carefree pre-war self," wrote his friend and colleague Isidor Isaac Rabi years later, the man who will also wonder in all seriousness why Sam Goudsmit never received a Nobel Prize for his trailblazing work in quantum physics.

Things were never quite right again between Sam Goudsmit and the man whom the Alsos mission regarded as the most important German suspect in the search for Hitler's atom bomb: Werner Heisenberg, a colleague of the same generation, and a downright genius. In Goudsmit's 1976 obituary of Heisenberg, he remembers how Heisenberg reacted in 1943 to the news that Goudsmit's parents had been deported. The older Dutch physicist and resistance figure Dirk Coster, a good friend of Goudsmit, wrote to his Berlin colleague Heisenberg with an urgent request for help with the issue. Heisenberg ultimately answered on February 16, 1943, with a letter to Coster. In it he wrote how hospitably the Goudsmits usually received German visitors to Ann Arbor, that Sam Goudsmit was a leading physicist, and that he worried about the fate of his parents: "I would find it regrettable if, for some reason unknown to me, his parents in the Netherlands should get into difficulties."'

These are warm words, but whether Heisenberg actually undertook something with the German authorities officially is highly doubtful. When his letter arrived in Groningen, Isaac and Marianne Goudsmit had already been murdered in Auschwitz. Thirty years later, Goudsmit was not even very bitter about it. "I doubt that anyone among the physicists I know would have acted better under those circumstances." His own feelings of guilt won out over rancor.

As Sam leaves his parents' house, walking over creaking floorboards and crunching glass, a boy out in the street looks intently at the man in the American uniform. The boy extends a hesitant hand and gives Goudsmit a small piece of orange fabric. Sam accepts it and, feeling moved, puts it in his billfold with the photo of Jaantje and their daughter, Esther, who are at home in America. It feels like a real wartime medal, even though it is just a child's gesture.

CHAPTER 1

NOTHING TO LOSE AS YET

Leiden, September 1927

D ecember 1963. An office in the Rockefeller Institute in New York. In effect it has been twilight all morning long, an icy wind is blowing, earlier in the week it snowed a bit. Almost forty years after it all began for Samuel Goudsmit, the historian and philosopher of science Thomas Kuhn leans over a large tape recorder on the table and pushes the record button. Right!

Kuhn, a professor at the University of California, is intrigued by the question of how science works. Like an anthropologist, he conducts interviews with leading scholars. He has become well-known for his book *The Structure of Scientific Revolutions* (1962). Now he is seated across from Goudsmit, head of the Physics Department at Brookhaven National Laboratory on Long Island. Kuhn gets up and looks at Goudsmit for a long time. The microphone is located between them. The reels are turning: in a yellow-lit opening on the recorder a red needle waits, motionless, for the first words.

After long urging by Kuhn, the two men have agreed to speak, on the record, about Goudsmit's youthful years in Leiden, and after that about his time in Ann Arbor, Michigan. How, as a student, aged twenty, he became fully caught up in the circle of great physicists such as Albert Einstein, Paul Ehrenfest, Wolfgang Pauli, Enrico Fermi, Hendrik Lorentz, and Niels Bohr, the builders of modern science. And how he, as a physics student, had known how to give that science a small push.

What Kuhn wants is an eyewitness account, as detailed as possible. Oral history, the story of the participants themselves. How did the great physicists of those early years work? Who influenced whom? Who knew whom? What did they say to each other? How did they obtain their positions? How did the new insights into particles and energy come about?

The men begin at the beginning. Tell me about your home, Kuhn says. He is a forty-year-old in an existentialist-looking turtleneck who as a young student took a course from Goudsmit. "Pretend for the moment that we're doing a total biography."

Goudsmit, a stocky man in his early sixties, is dressed in his customary impeccable suit. On the recording tapes, which are still kept at the American Institute of Physics in College Park, Maryland, near Washington, DC, he hesitates audibly. Biography, fine, you hear him think. But first he wants to get something off his chest about the European educational system that produced him. Unlike the United States, he says, in Europe not only the well-to-do go to university, but especially the clever. Origins do not matter. He himself, says Goudsmit, came from a family of small Jewish entrepreneurs in The Hague, and in fact it had been settled that he would go into commerce and industry like them. That he opted for science and the university was seen as a mistake in his family. His wealthy cousins poked fun at him. And if his older sister had not kept his parents from doing so, they would even have taken him out of secondary school and placed him in an apprenticeship with a commerce-oriented family member in France or the United States.

Only, in the end, Sam Goudsmit did not choose science. That is not the way it happened. It was the other way around: science chose Sam Goudsmit. Almost in spite of himself. And in any case, to his lifelong astonishment. Years later, in 1971, he addresses the fiftieth-anniversary convention of the Netherlands Physical Society in The Hague.

"What historians sometimes forget is how large a role in the sciences is played by coincidence and luck. Many careers and discoveries are a combination of circumstances."

In Goudsmit's case, that combination of circumstances has a name: that of his fellow student in Leiden, George Uhlenbeck. Born into a military family in Batavia, Netherlands East Indies (today Djakarta, Indonesia), a year older than Goudsmit, he began in the same small group of students supervised by Paul Ehrenfest, the famous Leiden professor of theoretical physics. Before that, Uhlenbeck had attended the HBS in The Hague, a form of secondary education that at the time did not yet allow university entry; this was limited to graduates of the gymnasium with its classical curriculum. Therefore, he initially registered in chemistry at the Technical University in Delft. The very same year the education law was amended, and he could enter Leiden with his HBS matriculation after all.

The transfer was welcome. The system in Delft had been extremely regimented, with daily lectures and long afternoons filled up with lab courses. In Leiden, he found, to his delight, not much actually happened. There were hardly any lectures, nobody looked after the students. There were five students in physics, among them the seventeen-year-old Sam Goudsmit, who seemed to know even less than the others about the purpose of studying. Sam and George were very different. One was an industrious, meticulous worker, the other a somewhat nonchalant, dreamy lad who waited for whatever turned up. Clever, but not particularly hardworking.

They met each other when they were second-year students in the Huygens Society, an upper-year debating club. Uhlenbeck was present when Goudsmit, on the urging of his fellow club members, gave a talk, not about nuclear spectra for once, but about Egyptian mathematics. That was the start of a lifelong fascination with antiquities and hieroglyphs; all his life Goudsmit will collect ancient Egyptian art objects, and at the end of his life he will leave his collection to a museum in Michigan.

In 1922, his fourth year, Uhlenbeck took a part-time job to ease his parents' financial worries. He taught at the Municipal Gymnasium in Leiden for twelve hours a week. He found the work to be dreadful, could not keep order, and was irritated by the minimal motivation and preparation of the pupils. The young men were not gripped by the physics that he tried to impart to them; classical languages and art were held in much higher esteem. His experience in the classroom would become a key motive for going to America in 1927. Anything was better than becoming a teacher "in the provinces" after finishing his studies.

At Leiden University, Ehrenfest is the only instructor who maintains a close link with the students. Every Wednesday evening from eight to ten, in the front room of his house on Witte Rozenstraat, he leads a colloquium in which the latest developments and publications are discussed. Sam Goudsmit is invariably present, although he is often barely back from his job as assistant in Pieter Zeeman's laboratory in Amsterdam. During one of those evenings in 1922, Ehrenfest asks his students whether anyone is interested in a small job in Rome: instructing the two sons of the Dutch ambassador to Italy, Jan Herman van Roijen, and his American wife.

"Nobody reacted besides me," Uhlenbeck remembered years later. He can use the money, and he gets the tutor's job in Rome. Ultimately, it will last three years. Formally, he continues to study physics in Leiden, admittedly from a distance and mostly on his own. He has little difficulty taking the exams during the Easter and summer holidays, while he is visiting his family in Holland.

At the outset he instructs the Van Roijen boys in mathematics and physics. In 1923, the older, Jan Herman Jr., goes to The Hague and successfully takes his final gymnasium exams. After studying law in Utrecht, Junior follows in his father's footsteps and enters the diplomatic service. During the war he serves as a cabinet member in the exiled Dutch government in London, and in 1950 he becomes the Dutch ambassador in Washington.

Uhlenbeck enjoys his stay in Italy and earns good money. On the advice of Ehrenfest, he seeks out the young sorcerer's apprentice Enrico Fermi, one of the top physicists of the twentieth century, in Rome. Fermi knows Leiden; he went there on a working visit to Ehrenfest in 1924. A famous photo made at that time depicts the professor with his young visitor and several students, among them Sam Goudsmit and also Jan Tinbergen, later an econometrician and the winner, in 1969, of the first-ever Nobel Prize in Economic Sciences. Uhlenbeck would later remember that Fermi was tickled pink to be contacted in Rome by one of Ehrenfest's Dutch students. "There was certainly, except for Fermi, no one who knew anything about physics."

The mood in Italy is gradually becoming more disturbed. In 1923, Uhlenbeck watches Mussolini's Blackshirts marching through Rome. The atmosphere points more and more to chaos and violence. Perhaps, Uhlenbeck decides, it is getting to be time to leave.

Ehrenfest again offers comfort. The Leiden professor calls his talented student back in 1925 to be his lab assistant. He asks his student Sam Goudsmit to brush up Uhlenbeck's knowledge. In Italy, George has largely missed the recent tempestuous developments in physics. Furthermore, he has become bored, has bit by bit begun to study Latin, has even formed a plan to study ancient history. When George Uhlenbeck reports in Leiden in June 1925, Ehrenfest is taken aback by his elegant, slender appearance, and his enthusiastic tales about classical antiquity. Ehrenfest infers that George is filling his head with the wrong ideas.

The intervention succeeds. Under Ehrenfest's gentle pressure, physics also chooses Uhlenbeck, whether he wants it or not. In the summer of 1925, Uhlenbeck and Goudsmit meet at the institute for a couple of hours several times a week, Sam lecturing casually and intuitively, George listening carefully and inquisitively, taking notes, making calculations.

Meanwhile Sam Goudsmit, twenty-three years young in 1925, is considered in Leiden to be an absolute expert in the area of nuclear spectra. In 1921 he had traveled to Tübingen for the first time, accompanying his father on a business trip. There he looked up the spectroscopist Friedrich Paschen, who received the youngster from Leiden not as a student but as a colleague and showed him "the famous helium line with the fine structure, which he had set up with an interferometer in his laboratory. I did not understand it. I didn't know what it was all about until I came back to Leiden." In Tübingen, Goudsmit found what he missed in Leiden: experiments. A couple of years later he returned to Germany for six months to learn the tricks of the trade in the laboratories.

Goudsmit knows all the literature about nuclear spectra, meets all the significant researchers in the field, knows all the theory that stands behind it, and besides, has an unerring feeling and absolute memory for the patterns in the light of the glowing atoms. The nuclear spectrum is his only passion, he says later. The rest of physics passes him by somewhat.

As Uhlenbeck and Goudsmit spend time together that summer of 1925, physics is in a state of great confusion. To be sure, in 1913 Niels Bohr finally discovered how the atom is probably put together, a sort of solar system of electrons moving in fixed orbits around a positively charged nucleus. If an electron jumps from one orbit to another, a fixed amount of energy is freed in the form of light of *one* particular color. Observed through a triangular glass prism, a pale line is the consequence. Bohr's model of the atom immediately has a stunning influence in physics. In one stroke it establishes the Danish physicist's name, because it neatly explains the light spectra of a number of elements.

But there are also problems with the model. It is good for describing simple atoms with few electrons such as hydrogen and helium, but not somewhat more complex atoms. And frankly mysterious is that nuclear spectra turn out to be less simple in detail that was thought, with split lines that sometimes shift in the proximity of magnets.

That there is some kind of order in the confusing reality is clear. Niels Bohr used *one* so-called quantum number n to indicate in which orbit an electron moves within an atom. After a lot of puzzling, the German Arnold Sommerfeld devises a system that, by means of two extra quantum amounts, m and k, enables a better description of the peculiar regularities of the nuclear spectra. Sommerfeld discovers which values of m and k are and are not possible with a given number n: the so-called selection rules, which resemble Kabbalistics more than real physics. What all these numbers express continues to be a riddle. But the formulas work and that is already worth noting. The atom, too, follows rules.

At the same time, exceptions and strange patterns remain. Right around the time that Goudsmit and Uhlenbeck begin to meet in the Leiden library, the discussions about this subject reach a peak. The problem is the so-called anomalous Zeeman effect, named after the Dutch Nobelist Pieter Zeeman, a professor in Amsterdam at that point. Zeeman received his prize in 1903 for discovering that lines in a nuclear spectrum are split by a magnetic field. That is the normal Zeeman effect, which is described by Sommerfeld's numerical conjuring. But there is an even more subtle Zeeman effect, with still stranger subdivisions of the light lines that do not adhere to the known frameworks.

Sommerfeld proposes a *fourth* quantum number to describe the situation. That works, too, but the rules of calculation seem completely ad hoc, and once more the question is: what is going on from the point of view of physics? The fourth quantum number suggests to physicists that there is a sort of internal rotation within the atom, in one direction or in the opposite. But what is actually rotating is the subject of heated discussion. The atom as a whole? The nucleus? The atom minus the outermost electron?

Around the same time, the Hamburg-based physicist Wolfgang Pauli is wrestling with the question of why atoms exist in the first place. As Niels Bohr conceives it, what prevents all the electrons in the atom from coagulating in the lowest energy orbit? Seen simply from the point of view of electricity, the idea is that the negatively charged electrons would just as soon huddle up close to the positively charged nucleus. Something prevents that, but what?

In the course of 1924, Pauli sees how he may be able to understand that phenomenon. The atom as a whole is not the important thing; what matters are the individual electrons in an atom. If these occur in two different situations, and the rule is that two electrons in exactly the same situation may never be present in the same atomic orbit, everything falls into place. Then it is suddenly clear why, in larger atoms, electrons occur in shells with very definite numbers.

In December 1924, Pauli writes an article about his exclusion principle and sends it to Niels Bohr and his brilliant right-hand man, the young German Werner Heisenberg. Both of them are in Copenhagen, the capital of modern physics at the time. Their reaction is unexpectedly critical. Conjuring with numbers without any real physics behind it, is their assessment. Why, for heaven's sake, should only *two* electrons occur in an otherwise identical orbit? That's the central question.

Evidently the time is not yet ripe for the necessary key insight. The German-born American physicist Ralph Kronig, who wants to work in Tübingen for a while, examines Pauli's rules of calculation and the idea that something is going on with the individual electrons. Kronig proposes that the electron turns clockwise *or* counterclockwise, but he gets no hearing from Pauli for what he thinks. Pauli has become cynical as a result of the rap over the knuckles he got from Bohr. Besides, Kronig has made an arithmetical error that means all his estimates come out too high by a factor of two. Pauli calls Kronig's suggestion *witzig*, imaginative, but has no confidence in it and leaves it for what it is.

As Goudsmit, in the Leiden library in 1925, tries to show his fellow student Uhlenbeck the way through the jumble of quantum numbers, nuclear theories, and spectral data, they know nothing about Kronig's idea of a spinning electron. Anyway, as a classically trained physicist, Uhlenbeck has a lot of difficulty with all the quantum wizardry and number divination, but in August that handicap suddenly becomes astonishingly useful. As Goudsmit tells him about Sommerfeld's four quantum numbers and Pauli's ideas about the two electrons in every orbit in the atom, everything falls into place for him. "But that means it has four degrees of freedom, something like a spin," he interrupts Goudsmit in his explanation. That's the way it always is in classical physics. Goudsmit looks at Uhlenbeck in some bewilderment and confesses that to him, as he later recalls, this "was not even a clear idea."

However, the initial confusion gives way to enthusiasm. If every electron exhibits an internal rotation, the entire outline of the four quantum numbers and Pauli's exclusion principle suddenly form a logical whole. Suddenly, the strange number code gains a real physical meaning. That afternoon, the two young physicists decide to write down their insight. They submit the result to their professor, Paul Ehrenfest, who is delighted by his students' high ambition. He immediately sends the article on to *Naturwissenschaften*, at that time the leading international journal in physics.

In November 1925, the short article *Ersetzung der Hypothese vom unmechanischen Zwang durch eine Forderung des inneren Verhaltens jedes einzelnen Elektrons* appears. At once the names of Goudsmit and Uhlenbeck are on everyone's lips in the world of physics. To their own surprise most of all, Goudsmit tells the historian Thomas Kuhn forty years later. "To me it wasn't so important. To me it was just another rule to understand and explain the complex system of spectra. That is all it meant to me. I never really understood its importance at the time."

The rotating electron may be a solution to many puzzles in the spectroscopy of atoms, but the idea also presents a huge problem. A charged object that spins so rapidly emits electro-magnetic waves and therefore quickly loses its energy. This raises the question: how can an electron continue to spin forever? And conversely there is the problem of the electron's magnetism, which seems to be so great that an electron is actually too small to hold it.

It is none other than the old theoretical physicist Hendrik Lorentz who raises questions, notably in discussions with Uhlenbeck. Lorentz was Ehrenfest's predecessor in Leiden and is now curator of the Teylers Foundation in Haarlem, but in physics he is still a force to be reckoned with. The young physicist suffers an attack of stage fright and goes to Ehrenfest to withdraw the article he and Goudsmit have written. He wants to think about it again, especially after hearing Lorentz's reservations. "Lorentz was, of course, the god nearly, also for Ehrenfest, and certainly for everyone in Holland—the absolute authority," Uhlenbeck comments later. Their supervisor Ehrenfest reacts more matter-of-factly. He has already sent the article to *Naturwissenschaften*. He assures Uhlenbeck that it is no problem if the idea of the spin is put down in black and white, even if it may not be completely correct. Don't worry about it, Ehrenfest reassures the two: "You are young, you can afford a foolish mistake." Ehrenfest thinks they have nothing to lose: "You have no reputation . . . , you have nothing to lose as yet." Goudsmit and Uhlenbeck get over their hesitation, on the assumption that their cheeky article will probably soon pass into obscurity. Who knows, it may turn out to be nonsense, may even be rejected.

But the opposite happens. The article appears in November, and three days later a letter arrives from the German physicist Werner Heisenberg, a well-known exponent of what is known as the German boys' physics: brilliant young fellows who are afraid of nothing when theoretical physics is at stake. He calls the idea of the spinning electron "interesting," but does ask what the two authors have done with the factor two. Goudsmit does not have the faintest idea what this factor two is; he finds such details less gripping. Later it will become evident that the theory of relativity is also involved with the electron's rotation in the atom. This brings in a factor of two.

In the light of the bigger history, this is just a detail. Theorists around the world embrace the idea of the twofold electron spin almost immediately, because with its use a series of questions about atomic spectra can be understood. In January, Sam travels to Copenhagen to continue work on the idea of the electron spin with the great nuclear theorist Niels Bohr. The two do not get along all that well, it becomes clear. Bohr is an abstract thinker who arranges his thoughts while talking, but who needs a strong conversation partner to keep him on track. Sam Goudsmit, twenty-four years old and little persuaded of his own theoretical capacities, is not the right man for the part.

In a later lecture, Goudsmit will say that the idea of the electron spin was, in fact, low-hanging fruit: "You must not forget that the quantum theory did not exist at that point. What we understood about atomic spectra was guesswork, a form of soothsaying with numbers. You guessed at formulas and connections. Initially nobody knew why they were there."

Later that spring, the physicist Walter Colby knocks on Paul Ehrenfest's door in Leiden. Colby is a professor at the University of Michigan and has formulated the plan of establishing a new theoretical group there. To this end, he is looking for young talent. Does Ehrenfest know anyone? To Ehrenfest the answer is obvious: Sam and George, the two talked-about young physicists who have little to lose as yet. They are the ideal candidates for the position. Ehrenfest offers the two to Colby before he has even talked with the young men themselves. There is a good reason for that in Sam's case: he is in Tübingen for a session with the old spectroscopist Paschen.

Because of this, when Sam gets the offer there is no opportunity to discuss it with his fiancée Jaantje and other members of his family. But more important are Sam's doubts about America itself. You go to the States when you are on the run from the law, or if you are an adventurer with few prospects at home, or so he thinks. At that moment Europe is the leader in science; America, dull and boring, is an uninteresting backwater. And in that Europe Sam Goudsmit counts as a rising star. What, in short, do the distant plains of Michigan have to offer him?

Meanwhile, back in Leiden, George Uhlenbeck takes a very different view of the matter. He really likes it at the university, but his prospects for afterward are decidedly not tempting. If he does not watch out, Uhlenbeck realizes, he will end up as a physics teacher at some municipal gymnasium or other. In his family, something like that is seen as the highest goal a university graduate can achieve, but Uhlenbeck has already taught for a while and has decided that he is not cut out for a career as a high-school teacher. For him, therefore, the offer of a professorship in Ann Arbor is just what the doctor ordered. He has no idea that Goudsmit, then in Germany, may possibly have doubts.

For his part, Goudsmit has no idea how decisive his supervisor Paul Ehrenfest is being. He at once assures the American that of course Sam will accept the offer, in spite of his hesitation. In the conversations with Kuhn in the 1960s, Goudsmit says little about this. Ehrenfest actually made the decision for him, he claims. He was all right with that. He was rather immature in those days, and Ehrenfest was like a father to him. Only after Ehrenfest's death in 1933 does Sam take more responsibility. Some of his letters to Ehrenfest should never be made public, Goudsmit says to Kuhn. They are too embarrassing.

In Leiden, Paul Ehrenfest has from the outset urged talent scout Colby under no circumstances to take just one of his talented youngsters. "Michigan is in the middle of the wilderness. You have to have at least two men who can talk with each other, and actually even more," he expresses his views at the time of the doctoral graduation of Goudsmit and Uhlenbeck in the summer of 1927. When, after a voyage via Liverpool and New York, the two Dutchmen arrive in Ann Arbor, the first group of theorists, among them also the German Otto Laporte and the American David Dennison, is complete. Enough talent to have a serious conversation, at least provisionally.

Hectic months have preceded the double graduation of Sam and George on July 8, 1927. First of all, Sam and Jaantje have gotten married in January in The Hague, in the presence of both sets of parents, but without much ceremony. There is no time for a honeymoon journey, because Sam has to keep working. When he becomes a professor in America he will need a doctoral degree and thus will have to write a dissertation that will enable him to graduate. To allow himself to concentrate he leaves for Copenhagen, where he finds a place in Niels Bohr's institute. He knows it well. In the spring of 1926 he tried to work with Bohr in further developing the theory of the electron spin, but in the end nothing much came from that.

This time Sam hardly speaks with Bohr. His dissertation, *Nuclear Model and Structure of the Spectra*, turns into a rush job, badly written and structured, as he later says himself. It is the pinnacle of Sam Goudsmit's almost obsessive interest in spectroscopy. This began back in secondary school with popular-science books and articles about starlight, given to him by his physics teacher Van Lohuizen. His teacher had also been a student of Ehrenfest's and was the one who introduced the above-average student to his former professor.

In fact, the student Goudsmit deals with nothing besides the pattern of lines of light-emitting atoms, the spectroscopy. Evidently he has an extraordinary feeling for the subject, but he neglects almost all the other subjects in physics in which a theorist should actually be at home as well. Those around him prove to be forgiving; his talent is clearly in the field of nuclear theory.

On September 3, 1927, the RMS Baltic, one of the many ocean liners that cross the ocean between Europe and the East Coast of the United States in those days, sails from Liverpool. Sam, Jaantje, George, and his wife, Else, have arrived in the English port from Amsterdam a few days earlier. Their names are right next to each other on the passenger list, the men registered initially as "teacher." Someone has crossed that out and changed it to the proud "professor." According to the lists, their young wives are simply "housewives." After a calm nine-day voyage, the ship arrives in New York. The physicist Robert Oppenheimer, later the scientific leader of the United States nuclear project, is waiting on the quay for the Dutch physicists. They know each other from Leiden, where the theorist Oppenheimer, generally seen as a mercurially clever physicist, once spent some time. He puts up the two Dutch couples in the not too showy Brevoort Hotel and shows them the city of New York.

The young foursome are amazed by what they see, especially in lively Brooklyn with its nightclubs, but in the meantime they also prepare themselves for the continuation of their journey to Michigan.

The two bright lights from Leiden will go to work in Chairman Harrison M. Randall's Department of Physics in Ann Arbor, a full day's journey west and north by train.

In Ann Arbor, Sam's eye is caught at once by an automobile that is being fixed in the lab. His spirits rise. In Leiden and also Amsterdam, practical work of that kind would be regarded as unscientific and therefore inferior. Here a healthy enthusiasm for the real world dominates. Sam remembers a little theory about the viscosity of oil mixtures that he drew up a few years ago and thinks that it may be possible to make money from it in car-crazed America. Nothing comes of his plan. Sam discovers that his idea has already been published and is also being used in the petroleum industry.

Ann Arbor, a sluggish provincial town on the prairie, exactly as Ehrenfest pictured it to them, is as disillusioning as the new work environment is exciting. Sam will always remember the first Christmas holiday in Ann Arbor as a downright horror. All the Americans have gone home, it is ice cold and snowing, and the newlyweds are living in a two-room apartment in a wooden building. They do not yet have a car. "Through one window you could see the hospital, through the other the cemetery," Sam later writes in a recollection-filled letter to his daughter, Esther. "There was no kitchen, just a gas burner for boiling water and making tea."*

When the downcast Goudsmit spots an advertisement in the British scientific journal *Nature* for a professor of physics in Egypt— warm and exotic—he applies at once. It turns out to be in vain. The twenty-five-year-old physics professor from a provincial town in America is not even invited for an interview. When spring arrives, life improves for the Goudsmits. But during the first few years, Sam keeps an eye out for opportunities in Europe.

From the outset, Sam's wife, Jaantje, twenty-four years old, has no objection to the adventure in the distant United States. In countless photos she seems to be having an extraordinarily good time in Michigan, always cheerful and dressed in the hip-hugging fashion of the twenties and thirties, with sheath dresses, gloves, and little hats. Her apprenticeship in the fashion houses of The Hague is clearly in evidence.

Sam, in fact, regards their stay in Michigan as temporary. He does not tell anyone at the time, but he actually wants to succeed the old Nobelist Pieter Zeeman in Amsterdam. As a student, Sam worked for a while as an assistant in his neoclassical laboratory on Plantage Muidergracht, three days a week. According to his own account, this was because Ehrenfest had seen that he was not a pure theorist but would rather get close to the experiments. Colleagues said that in those years Zeeman was a rather weary man in his sixties who enjoyed his scientific fame but came up with little that was new. "The plan was to wait in America until Zeeman died and then to succeed him," Goudsmit later says candidly.

In the end, Zeeman does not need to die. In 1939 he retires at seventy-four years old, four years before his death. His position as director and his professorship are opened up. At his departure, Zeeman expresses the wish to be succeeded by a Nobelist. But it is Goudsmit, once an assistant in Zeeman's lab, who is asked to fill the position. Sam ponders whether to return to Europe, but finally he decides not to, realizing how in the intervening years the old continent has fallen under the influence of fascism and anti-Semitism. Seen in hindsight, he writes to his daughter, Esther in a letter filled with memories and family history, the decision not to succeed Zeeman saved his life: "If I had gone to the Netherlands in 1939, I would almost certainly have met my end in Auschwitz." Of the entire Goudsmit family in the Netherlands, just one niece survives the Holocaust. His sister, Ro, flees to France during the war and soon after that lands in New York. There she will build up a new life with her husband, Henry L. Woudhuysen, a chemist from Brussels.

After two years in Ann Arbor, Sam and Jaantje return to the Netherlands for the first time. They are there for a family visit, but also because Zürich University has offered him a position as Erwin Schrödinger's successor. He goes there to discuss it, but ultimately it is Goudsmit himself who declines the honor. He will insist throughout his life that he is not a theorist at all, or at most haphazardly and on the basis of a lot of intuition. "I had a few tricks. But hardly any tools. I understood nothing of most of the dissertations that appeared. I wasn't even truly at home in the theory of relativity. The laws of Maxwell. I had no idea."

Upon his return to Ann Arbor, the University of Michigan gives him a hefty raise in pay to make sure he stays, and a guarantee that he will not have to teach first-year courses any longer. These had not been successes in the years just past, as Sam is too vague and at the same time too impatient to lecture to a class of beginners.

On the other hand, the course in advanced mechanics, which he teaches to upper-year students, is an agreeable tour of discovery for Goudsmit himself. Only in Michigan did he begin to study mathematics seriously, he later says to Thomas Kuhn: "Before then I was good at guessing the right formulas. Very intuitive; what I knew well were the spectra we were working on. But for the real work I needed someone like George [Uhlenbeck]."

The high points in the rather sleepy life in Michigan are the annual summer schools, when leading colleagues from Europe and the United

States come to devote themselves to their profession in a relaxed way. Before the 1920s, American physics was primarily focused on experimentation and practical questions. In Europe, physics in the age of Albert Einstein aimed more at theory and mathematical physics, the great questions of space, time, and matter. Harrison McAllister Randall, head of the Physics Department in Ann Arbor, wants that, too. Starting in 1924, he organizes summer gatherings where professors and students can catch up with the newest insights and developments.

With the help of his right-hand man Colby, Randall invites European lecturers, preferably prominent ones, to Michigan, where it is pleasant to spend time on the campus lawns in the summer. "I shall miss the fine warm weather very much, the swimming in Portage Lake, the malted milk at Miller's, the conversations with the physicists and the other summery matters," a participant writes when he fails to come one year. Niels Bohr does come, Enrico Fermi, Paul Ehrenfest, Heisenberg, and countless others; the only founder of physics who never comes is Einstein, who is professor in Berlin and already too famous to be able to accept every invitation.

Generally, there are about a hundred participants at the assemblies, which last from two to as long as eight weeks. They sleep in all the available rooms, from student residences to guest bedrooms in professors' homes. These are the marvelous years of quantum physics, and a sizable number of the roughly five hundred real experts come to Ann Arbor at some point in the period from 1926 through 1932. Sam regularly assumes part of the administration, and at the same time enjoys meeting his old European teachers and colleagues.

Among the guests in the pre-war years is the brilliant German physicist Werner Heisenberg, not yet thirty years old like Sam and nevertheless undisputed founder of quantum mechanics. Friends and enemies alike regard him as a genius. Goudsmit has known Heisenberg for years. On Ehrenfest's urging, he guided him around Leiden and The Hague in 1926, chatting in German. They dined at Sam's parental home, where they talked over dinner about a huge display of fireworks planned for Scheveningen that evening. The two young scientists had no time for the fireworks. They get on very well together, but they have to get back to Leiden for some serious work.

In Ann Arbor, Heisenberg stays with Sam and Jaantje on a number of occasions, for the last time in July 1939. On the return journey to Germany in August, Heisenberg is almost the only passenger on the SS Europa; whoever can leave the rumbling old continent at that moment is doing so. To go there is out of the question.

Goudsmit and Heisenberg's professional friendship will be sorely tested during the war years. Not only is Heisenberg determined to pilot German science though the war with as little damage as possible, he is also involved in the Nazis' nuclear research program, whose members Goudsmit will round up in the footsteps of the Allied armies in 1944–1945. They will finally meet again in Heidelberg in August 1945, Sam as victor, Heisenberg as leader of the defeated German physics profession.

CHAPTER 2

APPEALS AND ZEEMAN'S CHAIR

Michigan, 1933–1940

The telegram Sam Goudsmit sends on January 22, 1939, consists of barely two lines. It is addressed to the president-administrator of the Municipal University of Amsterdam, Willem de Vlugt, who is also the mayor of the city. The text is in capital letters. That is normal for a telegram, but it unintentionally makes the message more emphatic. "TO MY REGRET, WILL NOT BE ABLE TO ACCEPT EVENTUAL APPOINTMENT TO ZEEMAN VACANCY," states the message. A simple GOUDSMIT concludes the telegram. There is no note of thanks. No explanation. Even the period at the end of the sentence is missing.

What the reaction to Goudsmit's refusal was in Amsterdam is unknown, but Sam Goudsmit himself later often spoke about his decision at the beginning of 1939. In interviews and in the handwritten Goudsmit family history he wrote for his daughter, Esther, we constantly find the same sigh of relief. If at that time he had not decided to stay in America but had returned to Amsterdam with his wife and daughter, in all probability he would not have survived the war. In that case he, like his parents and innumerable other family members, would doubtless have been pulled along in the maelstrom of history. Then, just like tens of thousands of his Jewish compatriots, he would have been deported to the East, quite possibly via the Hollandse Schouwburg (Theater), which lay within a stone's throw of the Zeeman Laboratory in the capital city. He had no doubt that he would not have

47

survived, let alone that Jaantje and Esther would have made it. "The Krauts would have murdered all of us,"' he writes to his daughter without much ceremony.

He made the decision not to go in a flash the day before, on January 21, 1939, after a reminder letter from Amsterdam had landed in his office in Ann Arbor. The year before, Goudsmit had paid a visit to the Amsterdam mayor and university board member Willem de Vlugt to talk about the upcoming succession to the aged Nobelist Pieter Zeeman. Since 1923, Zeeman had been running his own spectroscopy laboratory on Plantage Muidergracht, specially built for him in the style of the Amsterdam School. It was right next to the older, more baroque-looking laboratory of Johannes Diederik van der Waals, another Nobelist from Amsterdam.

Goudsmit knows both Zeeman and his lab, located on the park canal, bordered by weeping willows. In his later student years and while he was working on his dissertation, in the mid-1920s, he worked there as an in-house theorist two days a week while based in Leiden. Every Wednesday evening, he took the train back to Leiden to attend Ehrenfest's evening colloquium, at the professor's house on Witte Rozenstraat. He always has to change gears, he tells friends, from the somewhat crude manners in Amsterdam, where off-color jokes are common, to the discriminating intellectual atmosphere in Leiden. Goudsmit hesitated a long time before finally deciding which of the two worlds suited him best.

The part-time appointment in Amsterdam gave Goudsmit a welcome financial compensation, but the relationship with Professor Zeeman never became optimal. The somewhat shy and dour Zeeman was often away on his travels, and his lab had had ceased to deliver pioneering new insights. Certainly not of the kind for which Zeeman and Hendrik Lorentz received the 1902 Nobel Prize. Both active in Leiden at the time, they had been the first to see and explain why magnetic fields can split the color lines of incandescent atoms.

In 1939, Zeeman is approaching the age of seventy-five, in those days the pensionable age for university professors, and according to both friend and foe, urgently ready for emeritus status. However, the old physicist refuses to depart so long as no suitable successor has been found to become scientific director of his lab. In an attempt to settle the matter, the university proposes Goudsmit's appointment. Not a Nobelist, of course. However, because of his discovery of the electron spin, he's definitely a big name, certainly in the area in which Zeeman has been active, atomic spectroscopy. Moreover, at that moment Goudsmit is not yet forty years old, which seems to offer a guarantee for new zest in the somewhat sleepy Amsterdam lab.

Succeeding Pieter Zeeman has been Sam's dream for the last five years. When the question appears on the agenda, in the second half of the 1930s, he discusses it openly with his pal Uhlenbeck, who has returned to the Netherlands to become a professor in Utrecht. There he succeeded the theorist Hendrik Kramers, who himself took the place of the deceased Paul Ehrenfest in Leiden. In September 1933, the fifty-three-year-old Ehrenfest, depressed and in despair because of his mentally handicapped son Wassik, ended his own life after shooting his son. Among physicists, the shock created by the tragedy was enormous.

In their letters during those years, Goudsmit and Uhlenbeck write extensively about the glorious prospect that Sam, too, may be returning to the Netherlands. After all, their two young families have been very close. In these discussions, old Zeeman's lab is very much in the picture, as both of them know very well.

But on January 21, 1939, holding in his hand the reminder letter from Amsterdam that urges a reply before February 6 and at the latest by February 15, "on which date the Board of Governors will meet," Goudsmit definitively decides not to go after all. After Hitler's 1933 seizure of power, Europe is no place to take a wife and child. There is menace in the air. Most of the European liners sail to New York full of passengers and return largely empty.

A few days after the telegram, *Het Handelsblad* notes in a brief news report that the world-famous Dutch physicist Samuel Goudsmit, currently in America, will not be accepting the prestigious professorship held by the Nobelist Pieter Zeeman. A footnote in history, perhaps, but one that leads Sam's sometime student friend Arend Joan Rutgers, a chemist at the University of Ghent, to send a relieved letter to Ann Arbor. "Thank God, I am so glad to hear that. More than once I wanted to write you: Goudsmit, you are crazy if you do that: settle within 150 km of the German border. Then I did not write because of your parents, but now I am glad you have decided it for yourself. Actually, it would be a good thing if you could now persuade your parents to go to America, the earlier the better, I believe." He himself, Rutgers writes, has urged his parents to leave Holland. "But it will probably be in vain."

The Leiden theoretical physicist Hendrik Kramers reacts with incomprehension and disappointment to the news that Goudsmit has declined Zeeman's professorship. In a letter to Goudsmit, a reply to one in which Sam explained why he would not be coming back to Europe, he says that he is "very unhappy" now that "you and George are lost to us." Uhlenbeck had left for the United States a short time before, worried about the political tensions in Europe. It put paid to a scheme hatched by Kramers, a personal master plan to offer professorships to good theorists in various Dutch universities. That plan has collapsed now that Utrecht has lost its man and Zeeman does not have a successor. Actually, Kramers's master plan was partly responsible for the collapse: in accordance with the plan, Ralph Kronig recently transferred from Utrecht to Delft. In the end it all makes no real difference. During the war, most universities are closed for one reason or another.

Kramers, a living legend among theorists, who worked for years as the top assistant of Niels Bohr, sounds amazingly apolitical in these difficult times, but his response is fair and well-considered. "We do agree with you both, of course. One of the things George can do so enormously well, keeping people together and inspiring them, was almost no use to him here (Utrecht); now he can once again follow his calling completely." And something similar applies to Goudsmit: "You would have done a lot, an awful lot of good in Amsterdam, but others would have benefited more from that than you yourself," Kramers notes in his minuscule hand.

Uhlenbeck's departure from Utrecht has gone as quickly and quietly as possible. All the Uhlenbeck family's household goods are packed in "a huge box of 10 cubic meters," are already sent to Detroit to arrive there in September. The family follows soon after. Initially Uhlenbeck goes back to work at the University of Michigan; later he finds a position at Columbia University. He will stay in America for the rest of his life, and in 1952 becomes an American citizen. In 1961, Uhlenbeck gets a professorship at New York's Rockefeller University, where he stays until he retires. At that time, Sam is working at the Brookhaven Laboratory on Long Island, not far off. For his work as publisher of the *Physical Review*, he often has to be in the city.

In 1964, Goudsmit and Uhlenbeck share the podium one last time and accept the prestigious Max Planck Medal in Berlin for their theoretical work, especially the discovery of the electron spin four decades earlier. After a Lorentz Medal from the Royal Dutch Academy of Sciences (1970), Uhlenbeck also receives the Wolf Prize in 1979, if anything an even greater scientific award. The year before, his dear friend Goudsmit suddenly died of a heart attack in Reno, Nevada. Out of friendship and respect, Uhlenbeck donates half of the Wolf Prize money to Sam's second wife Irene.

Since his visit to the Netherlands in the summer of 1938, Goudsmit has been engaged in difficult deliberations with his parents. He tries to get them to come to the United States, away from an ever-grimmer Europe. They hesitate. Will it be so bad? And what are they going to do in that distant America?

Half a year after Sam's decision not to return to the Netherlands, Hitler's troops invade Poland on September 1, 1939, and a new world war seems unavoidable. The Jewish calamity has been going on for over six years, although largely in Germany, which has been governed by the Nazis since 1933. The Netherlands has become almost used to the arrival of countless German refugees, mostly Jews who no longer have a life at home.

The situation causes Sam, far away in America, ever greater concern. Whenever possible he reads Dutch newspapers, and he follows the political developments in Europe closely. Every now and then he is in touch with the Dutch ex-pat community in New York, where a newspaper called *The Knickerbocker* is published specifically for Dutch people in America. He regularly signs petitions against Nazism and donates a few dollars to anti-Nazi campaigns in the United States.

In December 1937, he makes a telling decision: in a brief note to the treasurer of the *Deutsche Physikalische Gesellschaft* (DPG), Walter Schottky, he resigns his membership. He has been a member since he was in Germany as a student in the 1920s, and he has always retained his membership during his American years. He knows many German physicists well, and their professional association has long counted as a logical place for exchanging ideas and making contacts. In his archives you can find the annual requests for membership renewal. In recent years, he has been paying eight Reichsmarks annually.

To accompany his resignation, Sam chooses words that are incapable of being misunderstood: "I am disappointed that the Society has never protested as such against the bitter attacks made on some of its prominent members. Add to this that these days few contributions to physics emanate from Germany. The most important German exports are propaganda and hate."*

Schottky reacts with shock, especially because this involves the resignation of a foreign member: "External relations are of the utmost importance."*

At that time, the DPG, the professional society of German physicists, is chaired by another Dutchman, the physical chemist Peter Debye, a recent Nobelist who was born in Maastricht and made his career in Germany. He is also director of the Kaiser Wilhelm Institute for Physics in the leafy Berlin suburb of Dahlem. Debye, who has always retained his Dutch nationality, will be removed from his position in 1939 because the Ministry of War is initiating research into nuclear fission in his institute. But until that time, the Dutch scientist seems willing to make the best of a bad situation, particularly for the sake of science, in defiance of the suffocating German political atmosphere.

In 1938, Debye adds his signature to a letter accompanying the annual request for membership renewal which announces that "in view of the prevalent compelling circumstances" Jews are no longer welcome in the DPG. "In conformity with the direction I hereby call on all members who answer to this description to notify me of their resignation from the Society."* The letter is signed with a flamboyant "Heil Hitler" and "Debye, president."

The DPG has been going downhill for years, members are resigning, and especially Jewish scientists are leaving Germany. Sam is aware of this, and at the end of 1936 he discusses membership issues with Walther Gerlach, a former fellow student whom, at the end of the war, he will interrogate about his work on potential German nuclear weapons. Gerlach refers to him emphatically as "my old comrade and friend" and tries to keep him within the association. But Goudsmit wastes little time beating around the bush: "Sometimes I don't see the use of further support for the DPG. The inhuman treatment of many outstanding German scientists makes me deeply sad.'"

That "Heil Hitler" at the end of Debye's letter was standard in those days, but many years later it will give Debye's name a bad connotation. The biography *Einstein in the Netherlands*, written in Dutch and published in 2006, suggests that during the war Einstein, who had emigrated to America, wanted to have nothing to do with Debye because the Dutchman had collaborated with the Nazis.

On February 1, 1940, Debye arrives in New York by ship from Germany, officially for a lecture tour. In itself that is a striking move, soon after the outbreak of war in Europe. Debye's career in the Third Reich had the result that his arrival in the United States was watched with eagle eyes, notably by German-Jewish scientists who had to flee their fatherland. Was Debye not simply a spy? Or at least morally blameworthy? Above all the expulsion of Jewish colleagues from the DPG was held against him. After all, if he were sound, he could have protested by resigning as its president.

FBI reports from that period reveal that Sam Goudsmit, too, draws Peter Debye's background to the attention of the American authorities. In a letter, he reports that Debye, in his new position at Cornell University, could be in contact with sensitive defense projects. In September 1940, an FBI agent interrogates him about the letter. Earlier that year, Albert Einstein drew the attention of Cornell University and the FBI to a letter he received from a Swiss colleague, who claimed that Debye was personally acquainted with Nazi grandee Hermann Göring. On December 28, 1940, the decision is made not to admit the Dutch-German physicist to secret research anywhere in the United States. Only in 1944 is that restriction ended. The injunction does not keep Debye from working for the Bell electronics company in 1941 on the improvement of radar systems, and later on synthetic rubber.

In 1942, a remarkable interview with Peter Debye appears in *The Knickerbocker ("Free Netherlands") Weekly*, the wartime newspaper for Dutch ex-pats in the United States to which Sam also subscribes. The cover carries his photographic image, surrounded in the background by imposing high-tension installations. In a question-and-answer article, Debye speaks openly about the 1939 takeover of his institute in Berlin by the Nazi authorities to concentrate on work on nuclear fission, for explosives as well as new sources of energy. He also mentions that Werner Heisenberg is now the director. Aside from this, he declares himself urgently concerned with the American war effort. He himself is already part of a team involved in secret research on synthetic rubber, set into motion in a rush after the attack on Pearl Harbor and the start of the war with Japan, because the import of rubber from Asia has been choked off. That research, he says, is secret but is a matter of life and death. It may even be able to shorten the war, says Debye.

According to the historian Martijn Eickhoff of the Netherlands Institute of War Documentation (NIOD), Debye's emphatic openness about his work in America may have been intended to remove the mistrust that clung to him. The message was that Debye chose America. In the end, Goudsmit also acquiesces in this. After the war, in the 1950s, the two meet regularly in New York academic circles. The relationship is one of mutual respect, but a close friendship is out of the question.

The appearance in the Netherlands of a high-profile Einstein biography in 2006 briefly turned the Debye affair into an embarrassment, given that several scientific institutes had been named after the Dutch Nobelist. A lot of noise ensued, during which the institutes in Utrecht and Maastricht changed their names just to be on the safe side. Eventually studies carried out by NIOD and Berlin historians of science brought clarity. The question clearly was more nuanced than had first been thought. Debye, according to the final assessment, was not a collaborator but rather a totally apolitical opportunist who had but one goal: to pilot science and his Kaiser Wilhelm Institute through difficult times. The historical researchers found no evidence at all of personal anti-Semitism. Indeed, Debye had on a number of occasions assisted Jewish colleagues when they got into difficulties in Nazi Germany. Among other matters, he had been involved in the escape to Denmark of Lise Meitner, the woman physicist who had, with Otto Hahn and Fritz Strassmann, discovered nuclear fission in 1939. Another who helped her with this was Dirk Coster, a Dutch physicist, and a friend of Goudsmit and his parents.

During the pre-war years, Goudsmit regularly gets letters from colleagues in Europe as well as from total strangers, all trying to get work in America via him or in any case to obtain a letter of recommendation. One of those correspondents is Arend Joan Rutgers. Rutgers is a year younger than Sam and, like him, studied with Ehrenfest, under whose supervision he received his PhD in physical chemistry. He only got his doctorate in 1930, years after the tempestuous breakthrough of Goudsmit and Uhlenbeck, who have become professors in America. That year he spent some time in America himself, but it turned out he did not like it there. What is still intact is the friendship forged with Sam during their student days. In 1939, Uhlenbeck writes the previously mentioned letter in which he says he is glad that Sam has declined Zeeman's professorship.

The letter to Sam and Jeanne is not just a note from a relieved friend, however. Rutgers also makes abundantly clear that he himself wants to leave Europe, where ever-darker political clouds seem to be gathering. "For myself I would dearly love to accept a modest job in America, but it is probably too late," he writes. Still, would he not be able, he continues, to get a "small job" at three thousand dollars a year or thereabouts, in Sam's department or elsewhere in the United States? "Or is that too immodest? Yes, of course I made a bad impression nine years ago, when I did not much like it in the USA. But a person changes, and to be able to work quietly somewhere and enjoy my children now seems to me to be a paradise on earth. If you see a place, will you think of me?" He understands that it can be a "very delicate" business for one Dutchman to propose another Dutchman for a position. Perhaps, Rutgers concludes with an attempt at humor, he could be an asset for Rutgers University.

It is also evident from the letter that Rutgers has asked other physicists in the States for work: "Mayer and Oppenheimer wrote to me that it was difficult. I shall try writing to people who may have a direct interest in my work, such as [Irving] Langmuir." It all comes to nothing, Arend Rutgers stays in Belgium, gets through the war in relative peace and quiet, and eventually reaches the age of almost a hundred, known among professional colleagues and students as an author of physical chemistry manuals.

Rutgers's letter is by no means the only request for help with a departure from Europe that Sam Goudsmit receives. As early as the mid-thirties, letters arrive in the Dutch physicist's mailbox in Michigan, sometimes straightforward pleas for a job, or simply requests for a positive recommendation. At first Goudsmit reacts, however friendly his tone, in a rather cool and businesslike manner, or so it seems. If he does not know the people in question, he generally offers the advice to try somewhere else first, with someone they know. In later years, as the situation is clearly deteriorating, he is also willing to write friendly letters on behalf of strangers, as long as there is an indication that the researcher or student is of good quality.

Very occasionally Sam is approached by or on behalf of someone with a recognized name in science. In July 1938, Hans Hertz, son of the Göttingen mathematician Paul Hertz, writes to Goudsmit asking whether he can arrange a temporary appointment for his father, scion of a famous family of physicists. Hertz senior has been purged from Göttingen because of the anti-Jewish legislation and wants to come to America, where his son has already been working for years. He has no money. But he has known Lorentz and Ehrenfest, his son reports. In 1938, he gets an appointment at Yale; he dies in Philadelphia in March 1940.

On September 1, 1938, there is a handwritten letter from the famous Italian theorist Enrico Fermi, who at that time, without knowing it, is on the list of nominations for the Nobel Prize. Fermi is working in Rome, but he writes from Lugano, where he is holidaying with his wife, Laura, and the children. "Since the last time I wrote, so many things have changed so much that I have come to regret having turned down the position in Ann Arbor that you offered me. It is difficult to predict the direction in which matters will develop. But in spite of my natural optimism, I have to admit that I expect we are entering upon difficult years. In my case, the fact that my wife is Jewish can lead to unpleasantness for the children. I am writing this to you to indicate that I would very gladly accept a suitable position in America, if one exists. I would appreciate if you could let me know if something becomes available."

Fermi is referring to the race laws that have just been promulgated by Mussolini. In October he wins the Nobel Prize for physics for his theoretical work on nuclear fission. After traveling to Stockholm in December to receive the prize, Fermi and his family travel on to the United States, where he becomes a professor at Columbia University.

In the Goudsmit Archives, there is also an exchange of letters from 1940, made laborious because of language problems, with Professor Léon Brouillon, a leading French physicist who has fled to southern France after the French surrender in June 1940. Because of the pro-German Vichy government led by General Philippe Pétain, Brouillon wants to leave for the United States as soon as possible. He succeeds at the end of 1940, with sympathetic letters from Goudsmit among others in his briefcase. He gets to work at the University of Wisconsin, and later in the war participates in the American radar program.

One of the most remarkable requests for assistance is from a Dutch press photographer who would later become a well-known name in Dutch photo journalism: the twenty-two-year-old Sem Presser. On August 19, 1940, some months after the Netherlands was conquered by the Germans, he communicates by letter with "Prof. Goudsmit, Michigan University, MICHIGAN." The somewhat sketchy address, without mention of the USA, evidently has not interfered with delivery.

Since 1935, Presser has been a photographer for Press Service Foto Varia and has an office in his home at 137C Singel in Amsterdam, around the corner from big dailies such as *Het Handelsblad, De Tijd* and *De Volkskrant*. Sem is Jewish from top to toe, and in the years before the war he has been closely involved in the reception of German-Jewish refugees. He feels the coming of the Germans directly in his work and his life. He wants to get out as soon as possible and preferably to safe America.

Presser therefore registers for immigration with the American authorities and receives a so-called quota number. All aspiring immigrants get one. It indicates in which cohort they will eventually go to the United States. Presser has number 815, "very low, which leads me to assume that I am part of the first group of Netherlanders who will get permission to immigrate." Presser is in a hurry. A quota number like that is valid for only three months. "May I draw to your attention that matters like this cannot wait," he concludes his two typed pages of self-conscious supplication.

It is unclear how Presser hit on the idea of a letter to Goudsmit of all people, but his plan is well-considered and lucid. He wants to emigrate to the United States, he explains, and to that end there are two possibilities. The immigrant has to have capital, and Presser does not meet that condition. Alternatively, an American citizen, preferably a blood relative, has to make an affirmation of the immigrant's good character. Presser is asking for such an affirmation and hastens to say that he will pay the legal costs: "I shall regard their repayment as a debt of honor."

By way of reassurance, Presser explains to Goudsmit that he will almost certainly be able to provide for himself in the United States as a press photographer. And just in case, he is also prepared to take on other work. In the hope of being able to express his thanks to Goudsmit in person within the foreseeable future, he signs off with the utmost respect: "Samuel Presser."

Nowhere in the archives do we find an answer from Sam Goudsmit to the ambitious Dutch photographer. What is certain is that Presser did not come to America. In April 1940, he has taken a photo of Commander-in-Chief General Winkelman, conferring with Prince Bernhard in military headquarters in The Hague. It appears, among other places, in the popular American magazine *Life*.

After the general strike in February 1941, the only mass protest ever against the persecution of the Jews in occupied Europe, Presser finds a secure place for his archive and starts looking for a safe place for himself. He marries Ruth Schachno, a Jewish refugee from Germany. Together they go to Arnhem, where they go into hiding, and Sem is active in the Resistance under the pseudonym "Willem Jan Knol." Three times he is apprehended and three times he manages to escape. In 1944, he secretly photographs the English airborne landing, part of the Anglo-American Operation Market Garden, and the battle for the bridge across the Rhine at Arnhem. However, he loses track of the negatives after he buries them to keep them safe. After the Liberation, he learns that his parents were deported to Auschwitz during the last year of the war and were murdered immediately upon arrival there. Presser photographs the entry of the Allies into all the large Dutch cities. In 1946 and 1947, on assignment from the English media, he travels through post-war Germany, then still completely in ruins. After a long career as a well-known photographer and president of the Silver Camera, an organization that annually recognizes top news photos, Presser dies in 1986 in his birthplace of Amsterdam, sixty-nine years old.

Goudsmit sometimes receives insistent requests even from within America. The Polish physicist Roman Smoluchowski has reached the United States unharmed in 1940, "escaped from the terrible confusion in Eastern Europe, after a series of unusual and complicated adventures," but now he is looking for a way to quickly master the newest developments in physics.

Goudsmit knows Smoluchowski from a journey to Warsaw and Cracow, and he is pleasantly surprised that the Pole is in America. But he fears he cannot do much for him. Or wait, perhaps he can. He will mention his name at the experiment with X-ray crystals that is being put together in Ann Arbor. He himself will not be there that summer; he is occupied with business elsewhere. He does not mention that it is directly connected with the war.

At the same time, Sam is trying via his friends and colleagues to find out about the situation in the Netherlands, especially that of his parents. At the end of August 1940, M. L. Kobus, principal of the Princess Beatrix Lyceum in Flims, Switzerland, receives a letter from Ann Arbor. (The Lyceum is a sanatorium for seriously anemic Dutch children. The Swiss mountain air is believed to restore their health.) Kobus is a geologist Sam knows from his student days in Leiden. He has become a math teacher and finally a school principal. Kobus and Sam were both members of the Huygens Society around 1923. In student fashion, they still address each other by their last names.

Earlier that summer, Kobus asked Goudsmit whether there might be something he could do in America, as a teacher or possibly as a doctoral student, and how the educational system is put together over there. Sam answers him with a lengthy discussion of American schools and universities that ends with an expression of doubt that they are waiting for a Dutch geologist to show up. But then Goudsmit interrupts the page with a row of typed zeros and switches to a completely different subject: his own parents.

"I get very little news about my parents, they still live at the same address, I. Goudsmit, 137 Koninginnegracht, The Hague. They were on the verge of obtaining a visa for America, when the consulate in Rotterdam was destroyed. My father is without employment, and I am not allowed to send them money from here. Is it legally permitted to send money from Switzerland? If that is the case, would you be able and willing to act as intermediary for me?"

One thing he does know is that his sister, Ro, has fled with husband and child to unoccupied southern France. "They get assistance from the government and are fortunately very plucky and optimistic, and are getting along very nicely. It is a sad state of affairs, however."

Almost miraculously, Sam's scientific blood brother George Uhlenbeck made a lucky escape to the United States. He wrote to Sam, in a letter dated June 20, that in September they would be together in "good Ann Arbor town." And that does in fact come about. Until the last moment, Uhlenbeck has set and graded exams for his students, who will otherwise be without a professor to do this for them. On August 30, he and his family have set sail on the Queen Mary, which takes them to New York. The visas and boat tickets have been in their possession since the early spring of 1939.

A handful of people were aware of Uhlenbeck's plans for departure, among them his bosom friend Goudsmit, who heard the news from Uhlenbeck himself when he visited Ann Arbor at the end of 1938. One month later, Uhlenbeck thanked Sam for his hospitality and conveyed a terrible piece of news: the son of the late Paul Ehrenfest, also named Paul, died in an avalanche while skiing in the Alps. Just like his father, the young Ehrenfest was a physicist, and Uhlenbeck had just recommended him for a position at Harvard.

Uhlenbeck's letter of June 20, 1939, is important for another, more historical reason. In a postscript that comprises half of the letter, Uhlenbeck reports he has learned that Otto Hahn in Berlin has performed a new experiment with uranium. "It seems that with a neutron bombardment it can happen that the nucleus falls apart in two roughly equal parts. Fermi wants to test this by spreading uranium on the wall of an accelerator and then radiating it with neutrons." These are the first harbingers of the discovery of nuclear fission and especially the role of neutrons in that process. It is a matter of weeks before physicists begin to realize that something like this can lead to an unbridled chain reaction of nuclear fissions. With war in the air, the thought of new explosive material is news that attracts attention everywhere, within Germany as well as far away.

After the German invasion of Poland, Sam decides that his parents are no longer safe in Europe. As an American citizen (he had been naturalized at the end of the twenties), on December 11, 1939, he makes an official request for two visas. "Could you please send me the necessary forms, number 633, I believe, and let me know what information is required to obtain such a visa." He signs his letter with a simple "Samuel A. Goudsmit." Apparently, no one needs to know, in this context, that he is a world-famous physicist with contacts in Europe that will show themselves to be of priceless value during the war years.

Early in May, Goudsmit receives good news by way of the Immigration Service office in Detroit: the visas are complete and have been sent to the consulate in Rotterdam (a branch of the main Immigration Service office in the Netherlands, then on Keizersgracht in

Amsterdam). Goudsmit writes a week later to immigration official T. M. Ross, that he has received a letter from his parents indicating they will pick up the documents. But that letter from The Hague is mailed the day before the German invasion of the Netherlands on May 10, and since then all contact with the Netherlands has been broken. "I would therefore be pleased to learn, via you, whether my parents are in good health, whether they are still in touch with the American consulate, and whether they have received their visas." The outbreak of war may well have made all their savings worthless in one blow, Goudsmit realizes. "Is it possible to inform the consulate that I am prepared to pay all the costs of their journey?"'

The official would not have been able to do much. On May 14, the German *Luftwaffe* bombed out the central core of Rotterdam. In the fire storm that raged for days afterward, the American consulate on Vlasmarkt was spared. But the city was devastated and unreachable. What happened to the visas issued to Isaac Goudsmit and Marianne Goudsmit-Gompers has never become clear. In any case, they will never reach America.

On October 9, 1943, the Nobelist Pieter Zeeman, seventy-eight years old, dies in his home on Mauritskade in a cold and war-paralyzed Amsterdam. His successor at the university has still not been named. Zeeman's laboratory at 4 Plantage Muidergracht, with the beautiful stained-glass depictions of his famous spectral fissions above the doors, has been closed for years. The professor has not been there for a long time, and probably does not even know that occasional Resistance meetings take place there. He lies buried in Haarlem under a plain gravestone.

CHAPTER 3

THE SHADOW OF WAR

Boston, 1941–1944

O n March 20, 1947, the office manager and editor of the Master Reporting Company in Chicago, B. B. Calvin, writes a brief letter to Sam Goudsmit, who is then forty-five years old and has been a professor of physics at Northwestern University in Evanston, Illinois, for well over a year. The letter concerns the book they have been working on for some time, about Goudsmit's adventures in Europe during the last two years of the war. The manuscript is mostly ready, and what matters now are the finishing touches. Those have to be made quickly, but Goudsmit has not answered his phone for several days, and time is beginning to press.

Apparently Goudsmit went to New York for a conference, Calvin assumes. However, in view of the need for haste, he has taken the liberty of sending a carbon copy of a promised report to Sam by air express. All he needs to do is put a few things in the right places in the document file, so that a single entity will take shape that he can then edit at his convenience. Calvin trusts that the work has proceeded to the professor's satisfaction.

That report from Master Reporting is a verbatim account of a series of long conversations that Goudsmit has had at the beginning of March with the young New York journalist Edwin Seaver. On the weekend of March 8–9, 1947, they have talked about Goudsmit's war years, with interruptions for lunches and dinners together. A stenographer from the company has attended for all those hours in order to record everything. Seaver himself has made the occasional note. During the conversation, Goudsmit has walked over to his filing cabinet several times to extract documents, mostly letters from German physicists whom he has known personally for decades. There are also historical documents that he has obtained in Germany and other parts of Europe during the last year of the war. The papers he places on the table are mostly copies. The originals are in the possession of the US Army, as militarily sensitive material should be. On Goudsmit's desk there is a small metal cube, which feels uncommonly heavy to unsuspecting visitors. It is uranium that Sam has taken in 1945 from the German supplies in Munich. Quite harmless, he always explains to his visitors.

The transcript consists of 237 typed and numbered pages. A capitalized MR. SEAVER keeps asking the questions. DR. GOUDSMIT, also capitalized, answers them at length. Goudsmit stores the smudgy carbon copy of the text in his archive, which will be made public in its entirety by the American Institute of Physics years after his death. Also in the archive is a second carbon copy, another eighty pages long, of further conversations between the two men on March 22 and 23. These take place in New York. Goudsmit has received the first parcel and read through it, supplying corrections and explanatory notes where necessary.

In the second series of conversations, it is clear that the ice has been broken. Jokes are cracked from time to time, the tone is more relaxed than before. After fifteen hours of talk earlier that month, the men seem to be on the same level, no matter their differences, the professor and matter-of-fact war hero versus the eager journalist, in search of the truth and good stories.

The meetings of the two men are a preparation for a book with the simple title *Alsos* that is scheduled to be published in October 1947, by Henry Schuman Inc. in New York. The book tells the story, from the inside, of the secret hunt for the nuclear secrets of Hitler's Third Reich that Sam Goudsmit carried out in Europe between March 1944, and October 1945. In the wake of the Allied armies, sometimes even ahead of the front, Goudsmit and a group of scientists and military men went through universities, institutes, and hidden labs with one crucial question in mind: how close were the Germans to making an atom bomb?

The idea for *Alsos* comes from the publisher Henry Schuman, who in those days has his offices at 20 East 70th Street, very close to Central Park. After the war, Goudsmit has repeatedly participated in public discussions of nuclear weapons. That debate then comes to focus on the uncomfortable truth that the Americans knew as early as the beginning of 1945 that no German atom bomb existed, but that they nevertheless persisted with their own Manhattan Project, with the destruction of Hiroshima and Nagasaki as a consequence. Many physicists and engineers who have worked on the American bomb are frankly shocked that they were never informed of the German failure to develop a nuclear weapon.

At the end of 1945, Goudsmit speaks about the matter during Senate hearings, drawing on his experiences in Germany and Europe. Around that time, he is approached for the first time by the New York publisher, asking if he will write up his wartime adventures. There is a political issue here; nuclear weapons have become a controversial subject. Besides, Goudsmit's secret nuclear mission, his hunt for scientists in a war-ruined Europe, is an exciting adventure that could become a downright bestseller. It is a story people will enjoy.

Goudsmit is not much interested at first, not least because it is by no means clear whether he can speak and write freely. After all, much of what he came to know during the Alsos mission is nuclear in nature and, in fact, still secret information, certainly during the early days of the Cold War. But gradually he works out a usable rule of thumb governing what he can and cannot tell: technical details that have been in the newspapers are public and may be mentioned openly. A lot appears in the newspapers in those days; especially the *New York Times* is very well-informed and authoritative.

Aside from this, however, there is also the more commonplace issue of time. After years of war work in Cambridge and Boston, and more than a year on the warpath in Europe, Sam is glad simply to be home with his Jaantje and their thirteen-year-old daughter, Esther. He has left the University of Michigan for Northwestern University in Evanston, a northern suburb of the energetic metropolis of Chicago. There he lectures often and returns to his research into atomic spectra. A book requires a very different kind of discipline, and although he likes to tell stories and writes innumerable letters, putting together his story for a large reading public is an entirely different matter. "You clearly do not understand much about physicists," he admonishes Schuman. "A real physicist, someone like Fermi or Uhlenbeck, writes only for the *Physical Review* or *Nature*."

After some discussion via letters, Schuman proposes a compromise: he will supply a first-class journalist and writer who will turn Goudsmit's stories into a wonderful book. Sam will recount his stories orally and will get the final read-through and the last word on everything. Schuman proposes the New York journalist Edwin Seaver, later for many years the legendary book review editor of the *New Yorker*. After an initial meeting with the bright young reporter, the good-humored Goudsmit surrenders. Seaver knows nothing about physics or nuclear weapons, but it is abundantly clear he knows how to listen and write, and wants to know all the ins and outs of Goudsmit's life and work. From that moment, he becomes Sam's ghostwriter.

This takes place in complete silence. In none of the editions and printings of *Alsos* will the name Edwin Seaver ever be mentioned. Only in Goudsmit's personal archives and in those of Schuman, the publisher, can his name be found. When Sam Goudsmit holds a press conference at the time his book appears, on October 22, 1947, Seaver is not present. When asked about this, Schuman says the journalist is working hard on his first novel.

Sam Goudsmit's wartime adventure begins long before there is any consideration of an Alsos mission or secret reconnaissance in war-immersed Europe. On January 8, 1941, he receives a Western Union telegram in which his fellow physicist and good acquaintance Edwin Kemble of Harvard University asks him for help. The famous university has a shortage of professors, because its top scientists are increasingly being called away to research programs of the US Army or Navy. The question posed in the three-line telegram is: would Goudsmit consider a temporary appointment in the spring? The number of teaching hours is limited, Kemble writes; Goudsmit may choose the subject himself, and at $6,000 a year the emolument is not bad at all.

Goudsmit reacts positively at once; he is willing to help out his colleague Kemble. But he does make a few practical points. The University of Michigan has to be willing to grant him leave. His pension contributions have to be arranged. And if he comes, it will be without his family, and so the question is whether he can get a simple room. Moreover, he will have to go to Ann Arbor now and then, to visit Jaantje and Esther, and for his research.

In recent years, that research has shifted to the area of nuclear physics. There is a good reason for that. In 1939, the team of the German physicist Otto Hahn (with Otto Frisch and Lise Meitner) has demonstrated in experiments in Berlin that the nucleus in an atom of uranium can be split with neutrons; this discovery has led to feverish activity in the scientific world. Theorists and experimenters want to know everything about the new phenomenon. Furthermore, physicists around the world have realized at once the significance of the fact that the fission liberates further neutrons. It means that a chain reaction of nuclear fissions can be generated, even leading to an extremely violent explosion. The word "atom bomb" begins to be part of the buzz in the United States as well, far removed from Europe and a Germany in which Hitler has come to throw long shadows over society.

On January 22, 1941, with Europe at war, the University of Michigan gives him the green light: Sam gets his unpaid leave. At the beginning of February, he takes a modest room in the Commander Hotel in Cambridge, Massachusetts, a colossal six-story red-brick building in the center of the university town, for an indefinite period of time. The next day, he begins to teach his course in nuclear physics, meant for upper-year students in physics. Interest is modest, he mentions in letters written home. And the level of the students, he writes in evident satisfaction to Dean Barker in Ann Arbor, is not significantly higher than in Michigan. But Sam certainly likes it at Harvard. In a letter to Jaantje, whom he has begun to call Jay, and which is written in English, he reports on a rather wild spring celebration at the university, where alcohol is served. "Apparently the university does not care what happens to the girls." He himself went off the rails ("I had better not talk about that in a letter") and the following morning barely managed to get to his classroom on time: "At least, I seem to remember that." He will probably be at Harvard again later in the year, he concludes his letter, and will miss his family badly.

In July and August, he teaches in the Harvard summer school. The he travels to Long Island, where Jaantje and Esther are already vacationing. Together they return to Ann Arbor, to stay there until the term begins at Harvard and he has to go back to Cambridge.

Sam's substitute work at Harvard brings him in growing contact with fellow physicists at the famous Massachusetts Institute of Technology, on the coast a stone's throw east of Harvard. Goudsmit's reputation as a theorist is evidently no exaggeration. Within a couple of months, he is approached to join the research team working on radar systems at MIT. This new technology, in which radio waves are beamed around and the reflections used to locate objects on a map, is still in its infancy. When he is asked, Goudsmit initially hesitates. Radar is not exactly his field. In the course of the years he has lectured on all sorts of subjects, even classical mechanics, the subject he had tried to avoid at any cost while he studied in Leiden. That aversion, which was linked to his near-obsession with atomic spectroscopy, almost cost him taking his doctorate with Ehrenfest.

Besides, a short-term problem develops in connection with his formal admission to the military radar project at MIT. In principle, according to the security services, there is nothing the matter with the naturalized Netherlander. Yet it is not until the beginning of 1942 that he gets his clearance. He is not told what was holding up matters for all those months. Sam suspects that he is judged to be vulnerable because his parents still reside in German-occupied The Hague. Those worries continue to gnaw at him. Even in the letter to Jaantje about the wild spring festival, his thoughts stray toward his imperiled parents: "It has been a year since the Huns seized our Holland and I keep on thinking about that, in spite of alcohol and dancing."

In 1942, Goudsmit goes to work as theoretical advisor at MIT's Radiation Laboratory, known as RadLab. He is seldom home in Ann Arbor, lives in hotels and residences in Cambridge and Boston, austerely and with little more than a suitcase full of clothes. RadLab was established in 1940 by the physicist Lee Alvin DuBridge, a dynamic researcher and especially administrator, who would rise to being science advisor to Presidents Truman and Eisenhower, and later also Nixon.

Radar is a hot-button issue in the early years of the war in Europe. In the 1930s, not only the United States and Britain, but also Germany,

Russia, and Japan, have developed technologies for using radio waves to track down and locate distant objects. Most of the research and the current state of knowledge are still secret at that moment, but everyone knows there is a race to produce ultrahigh frequency radio waves. That will make smaller antennas and transmitters possible, but there is a catch: the system requires a very powerful source of microwaves. The Americans are working hard at this, but the necessary technology is invented in 1940 at the University of Birmingham in England: the magnetron.

British prime minister Winston Churchill realizes that in wartime it is wise to share this kind of invention with friendly nations. He seals a pact with President Roosevelt that provides for cooperation among American and British radar scientists. To that end, a lab is established at MIT under DuBridge's direction. Under the supervision of Vannevar Bush, head of the US Office of Scientific Research and Development (OSRD), the lab sets out to develop the technology further and especially to make it useful in the field, preferably also in airborne navigation systems. After the Japanese air attack on Pearl Harbor on December 7, 1941, the United States is at war and the RadLab is expanded rapidly. At its height, more than four thousand people are working at innumerable radar projects. Sam is one of them.

And the war work gives him a taste for more. From a distance, a shocked Goudsmit follows the way in which his old fatherland and the European continent are being trampled by the Germans. He considers what else he can do. In the summer of 1943, when the German advance has been halted and the war is stagnating, he gets an exciting idea.

In a July 1943 letter to Lee DuBridge, he makes a helpful proposal that will eventually turn his life upside down. Under the heading "Information about European physicists," he writes to DuBridge: "I have thought about a certain problem that will possibly come up in the not too distant future and in which I could be of service. Without doubt Europe will soon break up, probably into several parts. It would be advantageous to us to gather information as soon as possible about the work of scientists over there. As it happens, my personal relations with most of the physicists in Italy, France, Belgium, Holland and even Germany are very close. It would estimate that there are quite a few German physicists who still see me as their friend. Whatever the case, I have the feeling that my close relations could be useful in obtaining information that a commission of people unacquainted [with those scientists] would not obtain so easily." He concludes: "Should this question ever present itself, please keep me in mind.""

DuBridge's reaction to Goudsmit's proposal has been lost. But later that year, when the OSRD secretly approaches him, it becomes clear that the idea has not been lost. The OSRD, headed by the presidential science advisor Vannevar Bush, has been established in May 1941 to enable better coordination of research for military purposes. The OSRD has virtually unlimited means and reports directly to President Franklin Delano Roosevelt himself. It finances research in many areas, but focuses in deepest secrecy on what is called "Section S-1": the project in which thousands of scientists and engineers are developing the first atom bomb. After three years of unprecedented research activity in the Manhattan Project, the first atom bomb explodes in the New Mexico desert on July 16, 1945. In August, two American atom bombs destroy the Japanese cities of Hiroshima and Nagasaki.

But in 1943, that is still a long way off. Goudsmit is very much surprised by Vannevar Bush's visit. His original proposal to DuBridge was to interrogate European scientists about their activities as quickly as possible after the war. It did not occur to him that something like this might be desirable while the war is still in progress. In the March 1947 conversations with Edwin Seaver, he says he never thought about that sort of "espionage," as he calls it. But he travels to Washington for talks, and then it becomes clear what Bush wants from him: to find out what European scientists, Germans and others, are working on.

In theory, Bush is interested in all scientific information. But from the first briefings in early 1944, it is clear what the operation will really focus on: the question of what is happening in the area of nuclear fission. Are Hitler's scientists working to develop an atom bomb? That is the question and, for some time already, the fear.

The question about the potentiality of a German atom bomb does not come out of the blue. At that very moment, in the summer of 1943, the Americans themselves are involved in a project of that kind. Albert Einstein and Leó Szilárd, among others, have urgently warned the American president of the potential consequences of the 1939 discovery of nuclear fission in Germany. A bomb does not seem far off.

Officially, Goudsmit knows nothing about the American nuclear project. But by now, he knows fairly well what is going on. Colleagues and friends without number have dropped out of sight scientifically and have thrown themselves into secret research. Scientists like Goudsmit need only a hint to realize what it is all about: nuclear fission and the old fear that it will permit the construction of unprecedented explosives. "Everybody knew about it," he says to Edwin Seaver. "Among scientists you can't keep this sort of thing a secret." So not from enemy scientists, either? Seaver wants to know. "Not from the Germans either," Goudsmit says with conviction. "The only question was whether they had done something with it too."

Sam had already involved himself in a modest way with the question of what the Germans were up to in the nuclear area. In view of the German predilection for hierarchy, a German research program in that area could, in fact, be led by only one man: the brilliant founder of quantum mechanics, the theorist Werner Heisenberg. And Sam has known Heisenberg since his student days, when they walked together through Leiden and The Hague and dined at his parents' house on Koninginnegracht.

At the beginning of 1944, Vannevar Bush names Sam Goudsmit

as the head of a secret scientific mission in Europe. He continues to be a civilian and will be supported in his inquiry by a military unit. The assignment is, wherever the Allies advance, to search for laboratories and researchers, insofar as these are relevant to the question of whether something like a German nuclear project exists. The Americans, and with them the British, are quite convinced that it does. After all, it was the Germans who discovered nuclear fission, and the German physicists are regarded as leading the world. Add to this the German interest in advanced superweapons, and it has to be the case that Hitler is pursuing an atom bomb. It was precisely because of this specter that the Americans, after the warnings from Einstein and Szilárd about nuclear explosives, rig up their nuclear program. In December 1942, the émigré Italian physicist Enrico Fermi sets the first nuclear chain reaction in motion in a pile of uranium blocks under the bleachers of Stagg Field in Chicago. That achievement is the starting point of the Manhattan Project.

In his book *Alsos*, Goudsmit does not really comment on why he, of all people, is asked to lead the mission. Of course, it was because he knows, better than almost anyone else, German and other European scientists, many of them as colleagues or friends, and sometimes as fellow students. Add to this that he has never been actively involved in working on the atom bomb, so that he will not be a security risk if he is unexpectedly taken prisoner. He has only contributed to research on radar technology. At the same time, he is very well informed and completely able to assess research and researchers and to gather evidence.

What Bush does not know, but what makes the proposition even more attractive to Goudsmit, is his lifelong fascination with detective work. As a boy he spent hours at the door of a nearby police station in The Hague, dreaming about catching criminals. For a while he devoured detective novels, to the amusement of his family. And while he was a student, in his spare time he took a course in forensic techniques offered by an Amsterdam police officer. With the hunt for

German nuclear scientists, his fantasy is suddenly becoming reality. Sometimes, he will say later, the investigation in European labs and institutes was more exciting than the actual discoveries concerning the German nuclear program or the occasional proximity to the front.

In *Alsos*, he describes how more than once he is able to reconstruct secret documents and address lists from carbon copies in typing pools, the originals having been destroyed. In every abandoned laboratory or office they visit, he always first checks the mailbox for uncollected mail. "Erasing all traces of evidence at a crime scene is impossible," is his motto and conviction. In places like Paris, Strasbourg, Heidelberg, and Berlin, Goudsmit's men gather boxes full of secret documents and catch up to dozens of kingpins of the German nuclear research effort, however unimpressive it will finally come to seem.

After months of preparation in the War Department in Washington and an endless quantity of paper work, the physicist Sam Goudsmit, with a duffel bag and in civilian clothing, boards a military aircraft that will take him to London, the first stop in the biggest adventure of his life. In three days, the Allies will launch a huge invasion on the coast of Normandy. But the Alsos mission has already begun.

CHAPTER 4

HITLER'S ATOM BOMB

Paris, 1944

W hile on the Normandy coast the Allied invasion marking the beginning of the end of the Second World War is under way, an American military airplane lands at an air base just outside London. The sleepy passengers laboriously follow the path across the blacked-out air force base, clutching the required travel documents and saluting officers. The last early morning fog patches move across the field. Most of the passengers are in uniform, American officers and senior military staff who are going to lend additional support to Operation Overlord on the French coast. Physicist Sam Goudsmit, a stocky man in his forties with back-combed dark hair, dressed in civilian clothes, descends the steps and peers around for his contact. A somewhat older American colonel wearing round metal spectacles moves forward and shakes his hand enthusiastically. "Boris Pash; good that you've arrived, professor." It is June 6, 1944. The Allied invasion of the European continent has begun. And Sam Goudsmit is on the scene with a special assignment.

Sam and Colonel Pash have met before, during the preparation of the mission in Washington. They are of the same generation and like each other instinctively. At the same time, it is hard to imagine two men more different from each other. Goudsmit is a typical professor, learned, jovial, but also thoughtful. The muscled Pash is a professional soldier and former physical-education teacher, intelligent but above all energetic and afraid of nothing.

Earlier that year, in April, Goudsmit is working on radar technology at the Radiation Laboratory in Cambridge, Massachusetts, when he gets a phone call from presidential science advisor Vannevar Bush's Office of Scientific Research and Development (OSRD). He is asked to come to Washington for a conversation. Sam surmises that it is about a letter he sent to the OSRD sometime before, in which he proposed to enter into conversation with German scientists as soon as possible after the war. Not so much to pry their secrets out of them, Sam later says in an interview with his ghostwriter Seaver: "My idea was mainly that we would compare our notes and see how far each of us had gotten in physics during the war. I still assumed, very naively, that we had gone on working in much the same tempo. Perhaps men like Heisenberg and Schottky would have answers to things that we still had to find. And vice versa."

At the time, given the war situation, the military authorities considered Sam's idea to be unworldly and almost silly, but two years later they want to talk with Goudsmit after all. It was certain that the Dutchman had an interest in questions of scientific security. In 1942, he offered his wartime services and eventually he was called to MIT's RadLab. Even more significant, perhaps, was that in 1942 he passed on messages to the authorities about a couple of important German physicists who were going to travel to Switzerland for a conference. It seemed to him an excellent opportunity to get hold of information, and he proposed that he might go to the conference himself and cozy up to the Germans. Get to know the German scientists. Perhaps, he says, they will not be able to resist bragging about their wonderful work. The plan is realistic. Germany is not accessible for Americans, but Switzerland is a neutral country, and via the American embassy it should be possible for a scientist to travel there. In the end, the trip does not go through, and an Italian colleague who, on behalf of the Americans, tries to obtain information from the Germans gets nowhere. They really are not talking.

Sam's sharp-eyed attitude to the war has made some people take notice, and in April 1944, it brings him into the company of General Leslie Groves and his staff. Groves, a hard-driving United States Army Corps of Engineers officer, is in charge of the Manhattan Project. Officially, Goudsmit does not know this. Unofficially, leading physicists like him have known for some time what is going on in the field. Many of his scientific friends have suddenly gotten new appointments or have dropped out of sight altogether. Groves asks Goudsmit point-blank whether he is available for scientific investigative work. Not after the war and in America, but right now and in Europe while the Germans still have that continent firmly in their grip. That he is not officially a nuclear physicist is just as well.

Sam at once answers in the affirmative and learns that similar secret missions in Europe have already taken place, namely in Italy. American scientists have been pressed into service to examine installations in Rome and to interrogate Italian scientists about strategic German technology. These earlier missions have yielded little secret German knowledge, but they have provided a model for scientific reconnaissance operations in which military personnel and scientists advance together in recently liberated or conquered territory. Goudsmit and his team will need to draw on that experience as soon as the liberation of northern Europe begins.

The name the Americans have given the mission indicates a rather suggestive sense of humor. *Alsos* is the Greek word for *orchard* or *grove*. The name Alsos therefore refers almost directly to General Leslie Groves.

During the first weeks Goudsmit comes within an inch of being expelled from the mission before it has even begun. For his radar research at MIT he has excellent scientific contacts, with British colleagues at Oxford among other places. When he asks what they know about German advances in the field of new weapons systems, the military censor intercepts his message. Goudsmit gets rapped on the knuckles by Vannevar Bush's right-hand man for an ill-considered leaking of sensitive information.

Sam is genuinely surprised by the fuss. It seemed to him simply logical to ask the Brits what they know. "The belief at the time was that the Brits had a serious intelligence service." The matter finally fizzles out, and Sam is appointed as scientific head of the new Alsos mission. When he will actually leave for Europe is still unknown. That depends on the course of the war in Europe. What matters now is to prepare the mission and to draw up a list of useful experts who can be recruited for Alsos. The preparations take place in deepest secrecy.

In the spring of 1944, Sam can hardly have discussed the planned mission with Jaantje, but in their marriage something like that is no longer so strange. Sam has been working at the MIT Radiation Laboratory for more than a year, and once he even travels to a radar lab in England without her knowing about it. In letters to Ann Arbor, he sounds candidly cheerful and at the same time concerned about the well-being of his two "gals" back home. At that time Esther is ten years old. He was home on her tenth birthday. He gave her a xylophone, about which he later raises a complaint with the manufacturers. The sheet music of "Twinkle Twinkle Little Star" is full of mistakes, he grumbles.

For the rest of his life Goudsmit was amazed that he, of all people, was asked to assume the scientific leadership of the Alsos mission. A foreigner without any military experience, who in spite of his expertise in atomic spectra was not even a specialist in what the mission was pursuing. He was not part of the American bomb project; he was briefed about it only after he had accepted the assignment, alone among the scientists in the Alsos group.

But his position as outsider also had great advantages. If he were to be apprehended in Europe, he would never be able to betray crucial knowledge. Most important was that Goudsmit spoke German,

French, and Dutch, as well as English, and therefore could speak with scientists from those countries and could go through documents and archives in the search for evidence of nuclear research. The fact that he knew the presumed key players in a potential German nuclear program from earlier days, many of them personally, was seen as a dubious advantage at best. On the one hand, it would probably be simpler for him to sit down with old friends and really get to hear what was going on than it would be for a stranger to get the same result. On the other hand, the military-intelligence officers in the project thought that old friendships could be a problem. They might lead Goudsmit to go easier on detainees than was desirable. Later, he will have some difficulty with this himself. "I'm arresting men just like myself," he writes with some distress in a letter he sends home from Germany.

For Goudsmit, his briefing about the Manhattan Project in the spring of 1944 fell little short of being a farce. According to him, a lot of physicists had known for a long time that the project existed and also more or less who was working on it. Not because secrets were leaking out, but because the scientific community always knows what is being worked on, and because it is conspicuous when certain colleagues suddenly become less approachable. Goudsmit notices that people such as General Groves are very concerned about the ease with which the project has become known. If outsiders within the United States become wise to it so easily, are the Germans also aware that serious work is taking place on an atom bomb in America?

On June 3, 1944, a few days before D-day in Europe, Sam bids farewell to Jaantje and Esther at the New York airport. There is not much time. He has flown in from Washington and has to go on to London.

He is not carrying much luggage. In his billfold, he has a photo of his wife and daughter. In later letters home, he occasionally talks about Jaantje's elegant appearance. His wife writes back, but almost none of these letters arrive, as they are lost in the confusions of war. It will never become clear whether the censors are responsible or whether it is simply bad luck. All that autumn, Sam will complain bitterly about the lack of news from America: "Do write to me for a change; all the others get lots of letters, and you have no idea how important it is here to get messages from home."

He asks scientific friends whether they know if things are going well with his family and is cheered when that seems to be the case. When the astronomer Gerard Kuiper, already famous for the discovery of a distant belt of asteroids circling the sun, and a good friend of the Goudsmits, joins the Alsos team in Paris at the end of 1944, he puts Sam's mind at rest. At home, everything is going well with Jaantje and Esther.

As Goudsmit's plane takes off from New York, the British capital is in an uproar. At noon the BBC transmits the first official communiqué about Allied landings on the French coast. News reader John Snagge sounds understated as always, but the news is spectacular. The English prick up their ears. "This is London calling in the Home, Overseas and European Services of the BBC, and through United Nations Radio, Mediterranean, and this is John Snagge speaking. Supreme Headquarters, Allied Expeditionary Force, have just issued Communiqué Number One. Under the command of General Eisenhower, Allied naval forces, supported by strong air forces, began landing Allied armies this morning on the northern coast of France." For the sake of clarity, Snagge repeats the communiqué word for word.

The brief announcement follows an eventful night in which almost seven thousand Allied ships, from warships to landing craft, submarines, and merchant vessels, have steamed to the Normandy coast. In the landings on beaches code-named Utah, Omaha, Juno, Gold, and Sword, between Le Havre and Cherbourg, 156,000 troops have been put ashore by the end of the day in spite of German resistance. The losses are considerable: thousands of killed and wounded are lying on the beaches or floating in the water.

The term "D-day" is used for the first time in a BBC news bulletin at 1 p.m. Soon afterward, Prime Minister Winston Churchill addresses the House of Commons. Possibly to keep the House in suspense, he first offers an extensive report about the progress of the war in Italy, but then mentions that the invasion of Normandy has started. "So far the Commanders who are engaged report that everything is proceeding according to plan. And what a plan! This vast operation is undoubtedly the most complicated and difficult that has ever occurred." At that moment, the first infantry units are beginning to push inland.

As the day goes on, the German resistance in the coastal area stiffens. But it is too late. Initially, the German generals believe that the landings are a diversionary action and that the actual invasion will take place at Calais.

In Berlin, no one dares to rouse the sleeping Adolf Hitler before nine o'clock for a conference and new orders. And the commander of Army Group B in northwestern France, Field Marshal Erwin Rommel, is taking time off in southern Germany to celebrate his wife's fiftieth birthday. After a feverish telephone conference, he rushes back to France that afternoon.

Late that day, Sam Goudsmit is in a London taxi, on his way to Fourth US Army headquarters, where he has to report for further discussions with Pash and his men. Then he checks into the stately Mount Royal Hotel near Hyde Park. In the days that follow he and Pash, with

Pash's staff, start preparing a scientific reconnaissance mission. It must advance on the European continent in tandem with the advancing Allies. The operation's objective is clear: be the first to determine how far the Germans have gotten with the development of a nuclear weapon. Preferably before the British are there, is the unspoken intention of the Americans.

In the end, it will take more than two months before Goudsmit sets foot on the European continent. In his book about the Alsos mission, details of that kind are left unstated, but in a letter home dated December 10, 1944, he gives an extended and emotional account of his chaotic arrival on the old continent.

On Tuesday, August 29, Sam and his men drive into Paris. The city is elated, so soon after its liberation. "They know how to celebrate here," he writes in a letter to Jaantje and Esther a few weeks later. "Flags everywhere in the villages we passed along the way, the jeeps are buried under streamers. In the city, all the cafes open in spite of the blackout, full of entire families including small children, even at midnight. Immense numbers of people everywhere. They paid for our beer and the *café filtre*, the music was just like before the war, English and American music that they had not been able to listen to for years. We were just about the first Americans to enter this section, so we were stopped in the streets, children asked for our autographs, people invited us to dinner. It was a madhouse. And this had already been going on for a couple of days before we arrived."

In Paris he runs across a few old acquaintances, and a Jewish former student of his physicist colleague Kramers in Leiden who has been in hiding in the French capital for years. "What he told me gave me a bit of hope for the people who have been transported to Poland. I'm going to do everything to find out what has happened to our family and will probably not return home until I have discovered their fate."

On Monday, October 9, he writes a much less cheerful letter from Paris. He feels cold, wears his socks in bed, often wears his pajamas under his uniform while riding in the jeep so as to keep warm in the face of the coming winter. "This is my sixth letter to you. I still haven't gotten a single letter from you. You can't imagine how great it is to receive a nice letter from home. . . . Mail is more important than sleeping and eating. But at home probably you don't quite understand that. Perhaps Esther can write me a few lines on Sunday, the way she did from summer camp?"

Nothing much is happening with the Alsos mission at that moment; the Alsos crew are waiting for further troop movements, and there isn't much to do in Paris. "Sometimes I have a cognac in the café on the corner, but that's about it." Occasionally Sam attends the theater or goes to see a movie. "Mostly too awful for words."

Just as Sam is contemplating some story that will get him back to America, he finally gets mail from home, a long letter from Jaantje and a short one from Esther. "I almost sent you a telegram, but the mail arrived just in time," he telegraphs on October 22. Sam's mood is upbeat at once. "The city is wonderful, I've got a marvelous view from my office window. The streets look terrific, especially when the sun is out. That's all for today, my darlings."

A month later it is colder in Paris, and Goudsmit is complaining again. He is boiling water for a hot-water bottle while writing a short letter. "Part of my work is very, very nasty, and that depresses me. Meanwhile I'm a lot closer to the real war than a year ago." He cannot say much about his work. "It's important, a great responsibility. Occasionally it isn't so important and rather enjoyable, real 'mop-topping.' But even then it's work."

Luizebos, mop-top, is the nickname Professor Zeeman gave to his student assistant Goudsmit during his time in Amsterdam. For that reason, Jaantje sometimes affectionately calls her husband "Mop." If someone is fooling around, this is known in the Goudsmit household as "mop-topping." Very occasionally, Sam signs his wartime letters "*Luis*" as a token of affection.

On November 21, Sam complains to Jaantje long and loud about the absence of a good secretariat for Alsos. "They don't even get around to archiving." But there is excitement, too. "I fear the cold weather and I won't be able to write you for a while." Their good friend Gerard Kuiper, the astronomer, has also arrived in Paris; they will set out together. Some days earlier, Armistice Day has been celebrated with a long military parade, which Goudsmit and his men have gone to watch. Churchill and Eden and de Gaulle were in attendance; lots of North African soldiers paraded past, "with fezzes and modern weapons."

This time Sam ends his letter emotionally, with a declaration of love to Jaantje, whom he affectionately calls *Voutje*. "I close my eyes and imagine how you look. I long so much to be with you and to cuddle with you, because I love you and you are wonderful. But I wake up with a start, afraid of jealous scenes, unfounded accusations, that spoil everything again. Please! I long so much to return home to a warm and permanent haven. I've earned it, it seems to me. When Esther grows up she can't reproach me that I haven't done my duty in this war, I have made a real contribution and am still doing that."

He continues: "I'm grateful to be here, although it is tough and difficult and sometimes even dangerous. This is the greatest event of our century and no man would want to miss it. I'm grateful you never raised any objections to my wanting to go . . . I'm exhausted and sleepy. I shall dream that we're together and that I'm kissing you with real kisses. Nice and happy, above all very, very happy. Night, *Vousjepoes*, sleep tight."

Just over a week later, on November 30, Sam evidently has time for another letter, which with ten pages in minuscule handwriting is by far the longest wartime letter he will write to Jaantje. It is the only time that he describes his arrival on the European continent in detail; in his war book *Alsos*, much of his early wandering in France remains unmentioned.

After his arrival in London, Sam, in consultation with Boris Pash among others, prepared himself over a protracted period of time for the crossing to France and the hunt for German nuclear scientists that will follow it. Pash left for France a couple of weeks after the Normandy invasion. But Sam still had to arrange his gear, from a uniform to a helmet, blankets, and all kinds of equipment. On the evening of August 24, he excitedly packs his duffel bag in his hotel room near Hyde Park, aware that the real adventure is about to start.

At 7 a.m. on Friday, August 25, Goudsmit reports for duty with two American Alsos colleagues who have come with him from New York. They embark at a military air base, but then for several hours nothing happens because of a persistent fog. Seated on uncomfortable metal benches along the walls of the aircraft, they are able to cadge a couple of sandwiches from Red Cross workers who are with them on the plane.

Midway through the morning the fog lifts and the aircraft takes off. It flies low over the English Channel. Over the water, the passengers draw each other's attention to military convoys and the French coast that looms up in the distance. Sam realizes that things look more peaceful from the air than they are. When they get closer, villages along the French coast show serious damage from the heavy fighting of the weeks just past. The airplane lands at a temporary landing strip near Cherbourg, where it turns out no one is expecting them. There is no transportation. Sam and his travel companions commandeer a military vehicle that is waiting for a senior officer who has missed the flight. To begin with, they have to go to the local headquarters, a few kilometers away. In those headquarters, which are no more than a collection of tents and huts, there is no sign of Colonel Boris Pash, who has to set up the military mission. Nobody knows where he is. Possibly in the next city, several hours driving from Cherbourg. Goudsmit is assigned a truck with driver and sets out along roads which are clogged up with military travel.

Along the roads, they see notice boards and flags indicating mine-fields that have not yet been cleared. Driving between them, the Alsos team passes by abandoned German tanks, burned-out buses, cars, and even wooden carts, everything left behind by the hasty German retreat. It gives Sam a tremendous thrill. The Germans, "the Huns," as he calls them in his letters, have had a hard time of it during their flight. Farther on, many villages turn out to have been totally destroyed, especially where roads intersect, or bridges pass over rivers. But along the road they see people who wave and shout "*vive les Américains*," and make the Victory sign. The ill-at-ease Americans are growing into their role as liberators and hand out chewing gum and cigarettes.

For the first time, Goudsmit sees the horrors of war with his own eyes. Corpses of dead civilians are still being removed from ruins along the way. The image of a stretcher carrying a young woman stays with him. It is just as if she is sleeping, she looks so beautiful and peaceful.

It is getting to be late, and the question is whether they will reach the headquarters, given all the delays. They speak to an American soldier on a bicycle, who directs them to a nearby boys' school. There they are welcomed hospitably and served dinner. As he gets out of the car, Sam discovers that his duffel bag must have been left behind somehow. His blankets, clothes, shoes, equipment, camera, and helmet are all gone. As evening falls, they do manage to reach the next headquarters. There, too, there is no trace of Colonel Pash, nor of jeeps or Alsos soldiers. They spend the night in an empty girls' school. Exhausted, they sleep with the rumble of distant artillery in the background.

On Saturday morning, Sam decides to return to Cherbourg to sort out Pash's whereabouts and to get hold of a new outfit. Over the radio comes the news that General de Gaulle has entered Paris. The Germans have departed and have destroyed almost nothing, though it will later become known that the local commander refused to carry out orders. During the war, occupied Paris has become an excursion for countless numbers of German soldiers, which in a sense can even be seen as the discovery of modern mass tourism. The Germans have come to love Paris. Again, Goudsmit spends the night in an empty school. Shabbier this time, and among refugees. There is no water, the toilets stink dreadfully. Sam lies awake for a long time, sees how the distant sky is lit by explosions. He frets about what is expected from him and how ill prepared he is.

On Sunday, after trying for a long time, he reaches headquarters in Cherbourg by telephone. He receives orders to go to Paris and join Alsos there. He gets no official permission to use the truck that he and his companions commandeered earlier, but they take it anyway. Goudsmit is later cleared officially of the suspicion that he, as a civilian, has stolen a military vehicle.

Gradually it becomes clear that Pash's equipment and men have experienced a lot of delay in the logistic chaos of the weeks after D-day. Regrouping has to take place in Paris, the orders state. At the crack of dawn on Monday, August 28, 1944, Sam and his companions leave for Paris, a good day's journey away. It is gorgeous summer weather. The roads are in reasonable shape, so that even the primitive American truck with its wooden benches remains quite comfortable. The landscape changes. The villages they pass through show less damage. But along the road, the war is evident. Tanker trucks carrying fuel to Paris pass frequently. Coming from the opposite direction are open trucks filled with German POWs. Not much is left of the erstwhile master race, Sam thinks grimly. Here and there along the road, POW camps are being built, generally by the prisoners themselves. Conspicuously often they are being guarded by black American soldiers.

Closer to Paris, the situation once again becomes more gruesome. In a suburb, the railroad yard has been bombed. A locomotive, its boiler split open, stands next to a row of burned-out carriages. At the edge of deep bomb craters, the rails curl up; broken wires are hanging every which way. The surrounding houses and other buildings lie in ruins. This is the hard reality behind the maps and aerial photos that Sam has frequently studied during the preparation of the Alsos mission.

In the suburbs of Paris, people are at the side of the roads, cheering and waving with flowers and flags. Toward the inner city, it becomes quieter. The truck drives along the Boulevard St. Michel and Sam gets tears in his eyes. How often has he been in Paris in former days? And how undamaged it all looks. How very different London looked, pulverized and burned out by German bombers and V1s. Here everything is intact and open, the Jardin du Luxembourg, the Sorbonne, the Panthéon, Sainte Chapelle, Notre Dame, the Louvre. The bridges are still there, the Rue de Rivoli, the Place de la Concorde. But everywhere it is eerily calm.

In the American headquarters on the Champs Élysées, Goudsmit finally runs into Colonel Pash. He has already arranged for accommodation for Sam in an office in one of the imposing government buildings in the city. Jeeps, soldiers, and chauffeurs are also available. It is August 29, 1944; Alsos can get going.

After the fatiguing road to Paris, life there turns out to be very dull. Sam Goudsmit becomes part of the community of military men and intelligence officers whose task it is to search for Europe's nuclear scientists. But nothing much is happening, in fact. More and more of the Alsos men have adopted a strange pastime: they are growing mustaches. Sam does not care for this. The height of boredom, he calls it in a letter home. "Sometimes we play wild games of poker in our rooms. With French francs, because then it immediately looks as if lots and lots of money is at stake. But it has never been more than a few dollars."

To this point, Alsos is only a limited success, mostly concerned with intelligence work. Goudsmit is even disappointed visiting the famous physicist Frédéric Joliot-Curie (the son-in-law of Madame Curie) in his lab at the Sorbonne during the first few days after arriving in the capital. There are rumors that the Frenchman, an athletic man with a lightning-quick mind, made the cyclotron in his lab available to the Germans, but these turn out to be false. On the contrary, Joliot looked on with amusement as the Germans laid official claim to the equipment, but then did not really know what to do with it. In any case, they did not use it to pursue scientific ends. Plans to move the scientific equipment to Berlin were not carried out. In the last days of the occupation of Paris, the Resistance, with Joliot's full knowledge, used the lab to make Molotov cocktails.

Sam clearly expected a more adventure-filled role for the Alsos mission at this stage. Much of his work involves coordinating things, invariably from Paris, which after the first exuberant week of liberation turns out to be mostly a drab, exhausted city where the supply of electricity and fuel constitute a daily problem. In letters to Jaantje, Goudsmit complains that the American dollars one could use for some good shopping before the war turn out to have lost a good deal of their value.

With the liberation of Brussels early in September 1944, the daily routine is interrupted, really for the first time. Brussels is crucial in the international trade in uranium, which is mainly mined in the Belgian

Congo. Goudsmit pays a brief visit to the damaged Belgian capital. There he discovers documents belonging to the French mineral firm Terres Rares, which was taken over by the Germans and became part of the chemical firm Auer GmbH.

This company seems to play a role in a question that has already preoccupied the Alsos mission for months. Not long before the Normandy invasion, sources have confirmed, the Germans transferred Terres Rares's entire supply of thorium to Germany. Why this happened is unclear. Experts suggest that the radioactive ore could be used as the fissionable material for an atom bomb. The quantity of thorium that was removed is simply astonishing: hundreds of tons. That would suffice for several decades of normal industrial use. What do the Germans want to do with that much thorium?

This becomes clear when Alsos, after some investigation, tracks down a German chemist employed by Auer. He is apprehended in Eupen, a small town on the German-Belgian border near Aachen. The arrest is more or less a fluke. In Auer's Parisian office, a letter signed by a chemist named Petersen is found, addressed to a Fräulein (Miss) Wessel, a secretary employed by Auer who lives in Eupen. Petersen seems to have left for Germany some considerable time ago. But when Colonel Pash travels to the address in Eupen and rings the doorbell, Petersen himself opens the door. It turns out he has been having a secret relationship with the secretary. The couple is taken to Paris for interrogation. In the best tradition of German thoroughness, Petersen has a suitcase full of documents belonging to the company. "It was quite an event for us," Goudsmit notes with delight. "Here was our own first real Alsos Mission prisoner."

For several days, Sam, dressed for the occasion in military uniform, interrogates Petersen in a Paris hotel room. The results turn out to be disappointing, however. The man they initially hoped would know about the German nuclear effort from the inside seems to know hardly anything and is, in fact, not exactly bright. In the end, Goudsmit abandons the interrogation as a lost cause, and goes through the documents in Petersen's suitcase, page for page, at times while sitting in bed because the hotel in which they are staying is still not heated.

It becomes apparent that nothing in the matter of the exported thorium is what it seems to be. In various letters, different staff members of Auer discuss the post-war period, when the manufacture of war-related chemicals and gas masks will probably cease. They think that, seen in this light, perhaps thorium may be a good investment. One of the Auer employees is applying for a patent on a new kind of toothpaste enriched with radioactive thorium, claiming that it will whiten teeth. Even the advertising slogan has been drafted: "Toothpaste with thorium, for radiant white teeth." It is a distant echo of toothpaste with the radioactive element radium, which was on the market for a while before the war, until it became clear that the radiation did users no good.

In Petersen's suitcase Goudsmit finds a rail ticket for a trip to Hechingen in southern Germany. This destination sets alarm bells ringing. For months already, the rumor has been going around that leading German physicists have withdrawn to this part of the Reich, possibly because of their work on a German nuclear weapon. Somewhere under one of the Hohenzollern castles in the region, a large laboratory is supposed to have been established. Are Petersen, Terres Rares, or Auer involved in this in one way or another? Sam resumes the interrogation, but what logically seems to be a watertight case comes apart in his hands. Petersen claims that his mother lives in Hechingen, and that he has visited her as a private person. In the suitcase there is indeed a letter from his mother in Hechingen, in which she writes that parts of the town have become *Sperrgebiet*, blocked off, so that a visit will not be easy.

That, too, turns out to be less portentous than it seems to be, Petersen explains. The blocked-off parts of the town are intended for refugees from the bombed-out German cities and have nothing to do with possible laboratories or military installations. Petersen mentions that he was in Hechingen the year before. From his point of view, aside from the refugees nothing had changed in the year since.

A frustrated Goudsmit concludes that Alsos has been on a dead-end road for several weeks with Petersen. Only when Sam tells his ghostwriter Edwin Seaver about the toothpaste mystery in one of the interviews for *Alsos* does he seem to appreciate its tart humor. The story is presented in his book as an amusing anecdote.

On Wednesday, November 15, 1944, the Allies begin the campaign to retake Strasbourg, located on the Rhine, which marks the pre-1940 boundary with Germany. The fighting is heavy, and many buildings are damaged or even destroyed during the attack. The monumental university in the eastern part of the city is also hit, but here the damage turns out not to be too serious. The days are cold and foggy. At night, the occasional mortar shell falls on the city.

The stately old University of Strasbourg, founded in 1538, was evacuated rapidly at the time of the German occupation of Alsace in 1940. Equipment, archives, and staff were transferred within a few days to Clermont-Ferrand in unoccupied Vichy France. But the university was not closed. The German authorities decided to turn it into a scientific model institute for the new world order, and the institution remained open under the name of Reichsuniversität Strassburg, staffed entirely with Germans. Among them there are a number of leading German physicists who are on the list of Goudsmit's Alsos mission. One of them is the physicist Carl Friedrich von Weizsäcker, thirty-two years old at the time. Since the discovery of nuclear fission in 1939 in Berlin, Von Weizsäcker is a frequently cited nuclear physicist who often collaborates with Werner Heisenberg. Sam, ten years older, once met him in Germany, even though he remembers little of the meeting.

During the Allied actions in Strasbourg, the military members of the Alsos group, under the leadership of Colonel Boris Pash, leave directly from Paris for the front in eastern France. The plan is that they will locate the sought-after German nuclear scientists, their laboratories, and archives, as soon as they can reach the university. A tense Goudsmit remains behind in the French capital. Finally, after months of relatively slow intelligence work without much of a payoff, the real labor seems at be at hand. But Sam still has to wait for a message from Pash before the Alsos scientists can get to work.

When Strasbourg has fallen and Pash and his men reach the university in the eastern part of the city, they encounter very few employees. The physics department is deserted; the physicists have fled to the hospital and, disguised as medical staff, are trying to make their getaway. There is no trace of Von Weizsäcker and his colleague Eugen Haagen. They left for Germany weeks ago, their abandoned colleagues say disparagingly.

When Goudsmit, too, arrives in Strasbourg in a column of US Army jeeps, there is little grounds for optimism about the project. The marquee players have left, and the equipment he comes across in the laboratories hardly seems suited to real research into nuclear fission. There is a functioning synchrotron, a piece of apparatus for comparing atomic masses and measuring them, but for anything other than basic scientific research it is completely insignificant.

What does become clear from the piles of documents, reports, and letters is that Werner Heisenberg is indeed the central figure in German physics, and that he would certainly have been involved in any possible nuclear work. Moreover, his likely location becomes clear as well: southern Germany, far from Berlin. It is probably Hechingen; rumors have been heard for some time that a remarkably large number of physicists are staying there.

Hechingen has been in the picture ever since Sam was in contact with the British military information services, shortly after he landed in London to make preparations for the Alsos mission. The British were amused to find that the American evidently thought he was better informed than they were about the German situation. The opposite is true. That the Germans are interested in nuclear fission has been a certainty since the Allied discovery that they took over a heavy water plant in Norway. Heavy water is a variant of plain water, H_2O, in which deuterium D takes the place of ordinary hydrogen H. The fluid D_2O is important material for keeping control of a nuclear reaction in uranium. That the Germans have the use of the plant must mean that they are working on nuclear fission.

In addition, the British helped Niels Bohr, the world-famous Danish physicist, flee from his country before the Germans could take him captive. Bohr told the British that Heisenberg visited him at the end of 1943, and that he wanted to talk about the moral side of working on a nuclear weapon. After the war, there will be endless speculation over what has actually been discussed at that meeting in Copenhagen, but *one* thing is clear: the Germans were evidently at work on applications of the nuclear fission they themselves discovered in 1939. They were not very secretive about it, in fact; as early as 1942, German scientists such as Walther Gerlach in Berlin openly published a number of important elements in an ongoing nuclear reaction in uranium in an influential scientific journal.

The British became aware that when bombing raids on German cities such as Berlin and Hamburg became more intensive, laboratories and scientists were moved to southern Germany. British reconnaissance flights made aerial photos of suspicious activities in the area around Hechingen. New roads and high voltage wires were indeed installed, but gradually it became clear that these had nothing to do with a potential uranium project. What the British observed from the air were installations to exploit oil sands that were discovered in the region. The increasing scarcity of petroleum made their development an interesting proposition for the Germans.

When Goudsmit is flown to France in August 1944 and settles into recently liberated Paris, he already knows from British intelligence findings that Hechingen is likely to be of great importance to Alsos. In Hechingen, it is suspected, he will find Heisenberg, the undisputed leader of German nuclear science. The documents he sees in Strasbourg do make him aware that there may be less to be feared from the leader of German science than has hitherto been thought.

Goudsmit at once sees Strasbourg as the essential breakthrough of the Alsos mission; the most important work has now been completed. The Germans, he reports to Washington in November 1944, are approximately as far in their nuclear research as the Americans were in 1942, when Enrico Fermi got the first self-sustained nuclear reaction going in a pile of uranium blocks under the bleachers in Chicago. Perhaps the Germans have not even reached that point yet, Goudsmit thinks, after the discoveries in Strasbourg: "Everything is still at the academic stage. The conclusion is in any case clear: *there is no German atom bomb*."

At that very moment, in November 1944, people are working with might and main in New Mexico and Oak Ridge, Tennessee, to produce enough uranium for the first American atom bombs. The Manhattan Project, which will culminate with the bombs dropped on Hiroshima and Nagasaki in August 1945, is in full flight and not to be stopped. The first atom-bomb test ever will be carried out in Los Alamos on July 16, 1945. On that day, many of the scientists involved in the project are quite unaware it has been known for months that there are no indications of a German nuclear threat, although initially this was the most important reason for the Americans to undertake the project themselves.

In Washington, however, Vannevar Bush, Manhattan Project leader General Leslie Groves, and their staffs are not yet convinced at the end of 1944 that nothing is happening in Germany. True, in Strasbourg Sam Goudsmit and his experts have been unable to find evidence of a large-scale German uranium program, but that does not necessarily mean one does not exist. In any case, Strasbourg would not be the central place for a program of that kind. Cities in the heart of Germany, such as Berlin or possibly even Munich, are more likely candidates. Moreover, the archived material in Strasbourg could even be a diabolical red herring, intended to give Allied snoops the impression that nothing nuclear is going on in Germany.

Thanks to finding the archives, Goudsmit exposed the infrastructure of German science. But what was happening inside that structure still had to be determined. At the beginning of 1945, Sam is narrowly able to prevent the Americans from flattening Hechingen with bombs, just to be on the safe side. What finally decides the issue is that aerial photographs do not reveal a single nuclear installation. The Americans know how to recognize the industrial infrastructure of an actual uranium project better than anyone else, and they cancel the bombing raid. The idea that such a raid would seriously endanger not just thousands of refugees, but possibly also the leading German physicists, carries little weight with American military planners.

A horrifying by-product of the Alsos operation is the first concrete evidence that the Germans have been using prisoners for scientific experiments. Because the physicists have hidden themselves in the hospital and the university biology lab, Goudsmit and his team come across disturbing documents about such experiments. What mainly shocks Sam is the almost matter-of-fact way in which some researchers talk and write about their "material."

The central figure in the experiments is a German professor of anatomy: August Hirt. A member of Heinrich Himmler's SS elite corps, Hirt supplied various scientists with human guinea pigs. He himself performed experiments on live prisoners with mustard gas. He is also the man behind a chilling collection of "Jewish skeletons" that is found at Auschwitz after its liberation.

In November 1943, the virologist Eugen Haagen officially complained from Strasbourg to Hirt about the "material" supplied. Of the one hundred prisoners that had been sent, no fewer than eighteen had died during transport, and of the remainder only twelve were suitable for the intended experiments. "I therefore request that you send me another 100 prisoners, between 20 and 40 years of age, who are healthy and in a physical condition comparable to soldiers." This was the same Haagen who had done research at New York's Rockefeller Institute just a few years earlier, Goudsmit notes bitterly. Even then the German was already bad news: he was active in the German American Bund, a rabidly pro-Nazi movement in the United States. After the war, he will be sentenced for war crimes. In the late 1930s, Sam regularly signed petitions against the Bund and also donated money to its opponents.

When, after the success of Strasbourg, the dust has settled a bit, Sam finally has time for a long letter to his wife, Jaantje, and daughter, Esther. He sounds more cheerful than before, even if it is only because the heating in his temporary Paris quarters is finally working. There is even running water now, he writes on November 30, 1944. He does sound lonely, knowing he will not be home for the Christmas season this year. Goudsmit regularly plays poker with the military members of the Alsos mission. But his actual colleagues, the scientists who make up the core of Alsos, leave for London or even the United States as soon as their work is done. "They are never there when you really need them." As head of the mission, Goudsmit himself does not enjoy a similar freedom of movement; he has to stay. Moreover, he cannot be entirely open to Jaantje about his work, let alone about his suspicion that there is no serious German nuclear program in existence at all.

Nevertheless, he writes a full ten pages, until his eyes close with fatigue. To break the routine, he went with several colleagues to a show at the world-famous Folies Bergère night club. It was packed with noisy American servicemen. "It was grand, but a lot less wild than you may be inclined to think. You could have taken your old grandmother along to this show without a problem."

A few paragraphs later, Sam wishes he were a real writer: "There is so much to tell that I could write a book or play about it." At this point he does not yet have any notion of the incredible journey he will soon be making through Germany, devastated but resisting doggedly, on the road to Hitler's nuclear scientists and their leading light Heisenberg.

In spite of the Strasbourg success, Goudsmit occasionally finds his war work difficult. There is a ghastly side to his work, he writes to Jaantje in a somber letter dated December 10, "namely that I'm meeting people like myself, but from the other side. I'm glad that the men we were picking up had their families elsewhere. So I didn't have to separate them from their families."

He spent one night in Strasbourg in the empty house of an apprehended German physicist who had been taken to Paris. "Lots of toys like the ones you have, Esther. But also lots of Hitler stuff. It hurt me, I am too soft for this work, I believe. But at the same time they are so damned arrogant, these Germans."

Perhaps, he writes, the capture of the German research documents in Strasbourg means he will soon be called to Washington for consultation. In that case, there may be a Christmas at home after all. "That would be so beautiful. I could even bring a Nazi flag along, at least if you're looking for a scatter rug by the toilet. But I doubt you want that." Sam does not hide his homesickness for his family. "Cuddle with our Esther *for me*," he asks Jaantje in the concluding lines.

CHAPTER 5

URANIUM AND HEAVY WATER

Germany, 1944–1945

After grim winter-long skirmishing in France, Belgium, and the Netherlands, the American, British, and Canadian forces are on the left bank of the Rhine, poised to commence the conquest of the Germany heartland. The southern part of the Netherlands was liberated in the autumn of 1944. North of the Rhine, after the failed attempt to cross the river at Arnhem, the Germans were able to maintain themselves, and the Dutch population were starved and driven to despair during the frigid "Hunger Winter." In the cities, the last trees were felled, and combustible materials were stripped out of houses and buildings. In The Hague, the window sills and paneling of Sam's birth residence and of his mother's fashion boutique have bitten the dust.

Two weeks after the first breakthrough at Remagen, south of Bonn, on March 7, Allied troops cross the river in force on March 22. Upstream, the German resistance begins to collapse, weakened by continuous bombing raids and a lack of reinforcements. On Monday, March 26, units of the Seventh US Army occupy Mannheim, the city where the chemical firm IG Farben and the Daimler Benz engine factories were targets for intensive bombing for months. It is one of the centers of German war industry, where thousands of slave laborers and concentration-camp prisoners were put to work.

Two days after the conquest of heavily damaged Mannheim, Boris Pash and Sam Goudsmit, heading an Alsos convoy, cross the Rhine at

Ludwigshafen. Without incident they reach the historical university city of Heidelberg. Along the way they pass two destroyed American armored vehicles, still containing the bodies of the crew, surrounded by a silent group of villagers who are stunned by the violence of the battle. Sam will exclude the incident from his war book *Alsos*, just as he does with a lot of other gruesome details of his journey through a capitulating Germany.

One of the names on Goudsmit's Alsos list is that of a well-known German nuclear physicist: Walther Bothe. He has a lab in the Kaiser Wilhelm Institute for medical research in the city. From Sam the Americans know that, among other instruments, Bothe has a cyclotron, with which fundamental nuclear research can be carried out. Alsos, which spent the fall of 1944 investigating in France and the liberated parts of Belgium and the Netherlands, this time arrives perilously early. Not until twenty-four hours after Pash's specialists have taken over the laboratory does an American military unit show up with the assignment to secure the university lab.

Nervously, Sam enters the lab. He has to take Walther Bothe into custody and cross-examine him. "Here I was going to meet the first enemy scientist who knew me personally, a physicist who belonged to the inner circle of the German uranium project," Sam later states in *Alsos*. "It had not been too difficult to act authoritatively toward strangers, especially when I was backed up by a couple of real officers. But how could I be authoritative with Bothe, who was not only an old acquaintance and colleague, but certainly my superior as a physicist?"

After the breakthrough at Strasbourg in November 1944, it is not only crystal clear to Goudsmit that German scientists were not anywhere near the development of a nuclear weapon, but also, as he sees it, that they had never really thought about it. He makes an extensive report to General Groves and his staff in Washington, but he gets orders different from those he expects. The Americans are impressed by

Goudsmit's findings and the speed with which they have been made. But he has to keep on looking.

One nagging question lingers: what if it is all a brilliant piece of theater, the ignorance as a cloak intended to lead Goudsmit and his men down the garden path, while at the same time the Germans *are* working on a nuclear bomb? Or are there perhaps other German initiatives, more or less well hidden, to develop a bomb?

Add to this that part of Groves's staff does not fully trust Sam Goudsmit; the Dutch-born physicist knows personally most of the German scientists who would be able to run such a nuclear program. That makes him an ideal hunter of German nuclear scientists, but perhaps also less businesslike than the war situation demands. At the same time, it may make it easier to lead Goudsmit astray. After all, they are all scientists among themselves.

Goudsmit cannot understand the suspicions of the military men very well, but he acquiesces in his assignment. The war on the continent is still in full swing, the Allied forces are advancing into Germany. He has to determine on the spot whether there really has been no work on nuclear weapons. And if there has been such work, to track down any bombs. After all, Hitler and his henchmen do not waste any opportunity for boasting in their propaganda about the superweapons that the Third Reich still has in reserve. It is unclear what they are.

The physicist Walther Bothe, a short balding man in his fifties with a graying mustache, recognizes Sam Goudsmit at once when he is led before him by American soldiers. He greets him warmly. "Well, at last," sighs Bothe. Finally, someone with whom he can converse as an equal: a physicist instead of a soldier. He takes Goudsmit along for a tour of the lab, as a host who is receiving his guest. He shows the library, well-stocked with recent publications. And the cyclotron, at that point the only functional particle accelerator in all of Germany. Evidently Bothe does not know that at least twenty of these machines are operating in the United States.

Bothe and his team have done a remarkable amount of work during the war years, but a first impression indicates that almost all of it was pure research in physics. There can hardly have been time for applied war-related research. But when Goudsmit asks Bothe directly, he reacts guardedly. Formally, the German notes, he and Goudsmit are still at war. And so it cannot be expected from him that he speak about possible secret military research. He claims to have burned his secret observations, but Goudsmit does not believe that. A real scientist would never do something like that. But in closer interrogation, nothing points to work in nuclear physics. Bothe is finally set free.

Only after the German capitulation on May 8, 1945, does Bothe at last write a report about his own nuclear work during the war years. It contains little to surprise the Alsos team, who by this time, having made several detours on the way to Munich, have picked up and interrogated all the leading nuclear players. Bothe was important, to be sure, but he unintentionally hampered the German nuclear program in a major way. In 1941, he made a series of calculations that indicated that only a special, heavier form of hydrogen, deuterium, is suitable for the tempering of an enduring nuclear reaction. For some reason or other, he overlooked that the element carbon is physically much better suited for the process and, besides, is easily available in the form of graphite or ordinary coal.

This mistake means that the Germans began to show a conspicuous interest in heavy water. Early in the war, this alarmed the British and Americans. During a commando attack, the British destroyed a heavy water plant in Norway that fell into German hands not long before. Around the same time, the Americans, in their nuclear reactors, deployed carbon as the so-called moderator, the substance in the reaction that ensures that a chain reaction of splitting atoms will be controllable. Moreover, carbon is the key to the industrial production of plutonium, the actual core of the atom bomb. With their preference for heavy water, the Germans chose the wrong track at the beginning of the war. The reason for this was Bothe's faulty mathematics.

That Bothe, of all people, is working in the *medical* institute in Heidelberg is revealing. Earlier in the war he was pushed out of the physics institute when the university was taken over by the Nazis. Among the physicists, the big man at that moment was Philipp Lenard, a mediocre physicist who did get the Nobel Prize in 1905, but who had been a committed National Socialist for over two decades. Bothe, on the other hand, was reputed to be a moderate, not suitable for the education of fanatical students. He was still welcome at the Kaiser Wilhelm Institute for medical research in Heidelberg. In that institute, politics played a less important role.

In Heidelberg, the Alsos team moves into a couple of luxurious villas on Philosophenweg, near a *biergarten* in the upper town. During the next few months, the complex will function as the mission's headquarters for all of Germany, where resistance rapidly collapsed before the unrelenting advance of the Allies. One of the villas becomes the place where the captured German nuclear scientists are locked up. The building is hardly a prison: more like a resort.

Sam Goudsmit returns to Paris and resumes the coordinating administrative leadership of the Alsos mission. Now and then he flies to places where his expertise and sharp vision are required to assess new Alsos discoveries.

On April 12, accompanied by a couple of Alsos specialists, he flies on a military transport plane from Paris to Stadtilm in Thuringia, in east-central Germany. From the documents in Strasbourg and Heidelberg, it is evident that the physicist Walther Gerlach established his laboratory there. It was moved from Berlin-Dahlem in 1943 because of the ever-heavier Allied bombing raids.

The day before, units of General Patton's Third Army have taken the town of Stadtilm with breakneck speed, just a couple of hours after liberating Buchenwald concentration camp near Weimar. As he has done so often, Colonel Pash, Alsos's military commander, has advanced with the leading troops. In the area where wrecked armored vehicles block the roads, buildings are smouldering, and fallen Germans are still lying in the fields, he waits for Goudsmit and his men. He has already located the laboratory in a wing of a local school.

Pash describes the discoveries in his own book *The Alsos Mission*, published many years after the war. The lab is a shambles. Several families are holed up behind a large garage door made of oak. They claim to be sheltering from the violence outside. There are women and children among them, but Pash does not trust the situation, above all because they only reluctantly show their papers. Clearly digging has taken place in the school cellar; there is a pit that might have accommodated a primitive nuclear reactor.

Pash points at a few black briquets in a corner and asks the men what kind of material this is. Coal, they say, but when Pash picks up a briquet it turns out to be remarkably heavy. Reluctantly, the men admit that it is pressed uranium oxide. They begin to understand that the newly arrived Americans know exactly what they are looking for: the secret uranium research in which the men were participating. In the yard outside, another significant indication is found: blocks of paraffin that can be used as protection against radiation in nuclear experiments. Evidently they have been dumped in all haste.

Elsewhere in the old school building, the Alsos experts find a research lab that is still relatively intact, though almost pitifully small. "It was all on the scale of a rather poor university and not of a serious atomic energy project," Goudsmit later notes in his account.

The uranium that has been found is left over from the material the physicists used in an attempt to build a reactor from piled-up blocks of uranium oxide in a basin filled with heavy water. The pit in the school cellar is finished, but it seems never to have been used. Next to it is a second pit, with a big lead cover, in which there are traces of radium. Most of the equipment has disappeared from the laboratory.

The two research leaders of the lab, Walther Gerlach of Munich University and Kurt Diebner of the Kaiser Wilhelm Institute for Physics in Berlin, have vanished without a trace. According to those left behind, Gerlach left weeks ago; Diebner, on the other hand, left only two days ago, just before the surrender to the Allies. He took most of the uranium in a special Gestapo transport, Goudsmit gathers from the interrogations. Probably the stuff went in the direction of Munich in the south, one of the last German bulwarks and the area in which it is suspected that Heisenberg and his researchers are staying.

In Stadtilm, Goudsmit and Pash realize for the first time that the Germans definitely did attach a lot of value to research into nuclear fission and wanted to make every effort to build a bomb. "It had the full attention of the Nazi authorities. They must have expected a lot from it. It also refutes the statement often heard that the scientists had no support from the government. It was the scientists' lack of vision that prevented them from asking for more support, which they certainly would have gotten."

During the next few days the nuclear investigators find accounts in the archives Gerlach left behind in Stadtilm, which reveal once again that the German effort to produce nuclear fission is, in fact, still in its infancy. Above all, the documents are a revelation, because they are the first to contain the names of participating researchers, institutes, and financiers, as well as research budgets. All at once, the entire German uranium program is brought into sharp focus, a satisfied Goudsmit notes in his report.

Beyond that, Stadtilm is mainly a depressing place, cold and wet, without heat or electricity, full of refugees who are plundering wine cellars, wandering around drunkenly and looking for fights. The house in which the Alsos men are temporarily staying is crammed full of hunting trophies that cast strange shadows in the light of the gas lamps that illuminate the poker games. On the first night, Sam finds a cold plucked goose in his sleeping bag. The practical joker never identifies himself. It annoys him in spite of everything, and he complains about it in his letters home, written from the much more comfortable Paris.

On April 14, 1945, the Alsos mission leaves Stadtilm and returns to Heidelberg. Sam flies to Paris with a treasure trove of documents and materials that will be subjected to closer analysis. After that, he leaves once more for Alsos field headquarters in Heidelberg.

That same April day, another one of the Alsos leaders, the British chemist Michael Perrin, makes an important discovery much farther east, during an action in Stassfurt: in caves in the area between Magdeburg and Leipzig, virtually all of the uranium ore that the Germans took from Belgium in 1940 turns up. It is an estimated 1,100 tons, more than a million kilos, largely originating in the Belgian Congo. It is almost the entire supply of uranium existing in Europe at the start of the war. Almost, because 1,200 tons were taken to Germany in 1940. So, 100 tons are still unaccounted for.

In Washington, General Groves takes enthusiastic notice of the messages from Europe. It seems, he writes in a memorandum to British Field Marshal Montgomery, that the discoveries definitively establish that the Germans will not be using an atom bomb. "At least not during this war."[*]

Part of the uranium is taken to Antwerp by train and then shipped to the United States. Another part is transported by trucks to Hamburg, whence it is flown to England in a big transport plane. Michael Perrin, the man who tracked down the material, will lead the British atom-bomb program for a period of time in the post-war years.

One of the locations that play a central role in Walther Gerlach's Stadtilm documents is the small town of Celle, just north of Hanover. According to reports, in that north German town, the physicist Wilhelm Groth led research into the separation of fissionable uranium-235 from natural uranium ore with the help of super-rapid ultracentrifuges.

In the middle of April, Celle is in German territory that has not yet been conquered, so still beyond the reach of Alsos. Goudsmit travels to Göttingen, where a new temporary Alsos headquarters has been established in the former Reich Research Institution. Alsos scored a big coup in Göttingen with the arrest of Werner Osenberg, leader of the Nazis' defense research, and his entire staff. Osenberg carried a list of hundreds of names of Nazi scientists. That list of names will play a key role in Operation Paperclip, the American program in which countless German scientists are taken to the United States, for the rocket program among others. Also apprehended was one of the top men—Goudsmit does not mention his name—in Hermann Göring's feared air force research, the brains behind the first jet aircraft and long-distance rockets. He happens to be in Göttingen for a work-related visit when the Americans show up.

When Celle comes into Allied hands on April 12, 1945, Goudsmit immediately leaves Göttingen in a jeep with Robert Furman, his permanent military escort, who is also a civil engineer and chief of information for Alsos. In his book, Sam always refers to Furman as the "Mysterious Major," probably in response to Furman's own request. Two other Alsos men have already preceded them to Celle: their colleagues Walter Colby and Charles Smyth, who after the war will become the drafters of the famous secret Smyth Report about the extraction of nuclear energy, are waiting for them.

Along the way, Goudsmit and his major once again pass through a landscape of burning woods and wrecked tanks and trucks. The roads are jammed with British army vehicles and tanks, and refugees who are carrying what is left of their possessions on bicycles and hand carts. There are also foreign forced laborers who are trying to get home. High above them, the last dog fights are taking place, and shells land regularly without it being clear who is firing them. The road to Celle is an unalloyed nightmare for Sam. "There were men and women and a few children; they all wore rags and looked exhausted, but they sang and displayed their national flags proudly." To his surprise, an occasional refugee turns out to speak Dutch.

In Celle, they find the lab they are looking for in a former parachute factory. Goudsmit is greeted by an old acquaintance, the experimental physicist Wilhelm Groth from Hamburg. He knows Groth from when Sam studied in Germany for a few months on an American scholarship. That was in 1926, and Groth and Goudsmit became friends at that time, in part because they shared an enthusiasm for the author and humanist Thomas Mann. This, Sam thinks, may have guarded his old friend Wilhelm (he calls him Will) from succumbing to Nazism.

But he also knows that Groth is a patriotic German. With his colleague, the chemist Paul Harteck, at the beginning of the war he alerted the German authorities to the possibility that an extremely powerful explosive could be built with uranium. That letter, which Goudsmit has found in the archives, is comparable to the one that Einstein and Szilárd sent to the American president, Franklin Roosevelt, in 1939, which led the United States to initiate the Manhattan Project. Groth's epistle is a very clear incentive to take to take the nuclear fission so very recently discovered in Germany seriously, as a source of energy and as a weapon.

The meeting with Groth is uncomfortable. He and Goudsmit talk a bit about their families, but on both sides of the table an awareness dominates that the war has made them into enemies and has erased the old friendship. The lab in Celle consists of no more than a few rooms and is hard to take seriously. There are a couple of rapidly turning cylinders as tall as a man. Uranium vapor can be whirled around in them until the somewhat lighter and fissionable uranium-235 is separated from the heavier uranium-238. The Americans realize at once that uranium-235 was never obtained in significant quantities with the equipment. At most, some lightly enriched uranium was made for tests in a nuclear reactor with heavy water. In any case, the lab is out of commission now. Electricity has been lacking for months, Groth says.

From the military point of view, Celle is completely uninteresting. The technologies used by Groth and his associates will become important only after the war and form the basis for modern ultracentrifuges for the enrichment of uranium. At the beginning of the 1950s, the headstrong Dutch nuclear physicist Jacob Kistemaker, who heads an isotope lab in Amsterdam, more or less by accident attends a colloquium in Hamburg where Groth is speaking about his work. Eventually that will lead to the founding in Almelo of Urenco, which enriches uranium for fuel rods in civilian nuclear reactors. In the 1970s, Kistemaker will become the central figure in a controversy in the Netherlands because of his wartime work for Cellastic in Paris, a front for a German technical company. That Urenco's ultracentrifuge comes from Germany really does not help.

In the end, Celle becomes a historical footnote. For Alsos and Goudsmit, the key work is waiting to be done elsewhere. From the accounts they hear in Stadtilm and Heidelberg, Alsos's investigators have come to realize that they have to go to southern Germany, to the Hechingen-Bisingen region south of Stuttgart and Tübingen. The Americans know that Werner Heisenberg has a lab there. According to sources, Otto Hahn, one of the discoverers of nuclear fission in 1939, is with a group of close associates in the village of Tailfingen, close to Hechingen. Both places are still in German-controlled territory, so for the time being out of the reach of the Allied nuclear hunters.

The most important research locations in Germany have long ceased to be a secret. The letterheads of the Heisenberg and Hahn groups not only state openly that they are under the command of the Assistant Reich Marshal for Nuclear Matters, but also carry their addresses and even telephone numbers. It is no coincidence that the German physicists landed in the hilly area south of Tübingen. Walther Gerlach, head of the nuclear experiments, studied in that town, and he was also a professor there. He knew the surroundings well and thought that the narrow limestone valley of the Eyach River would provide sufficient protection against possible air attacks.

A bigger complication for the Americans is that French units will probably be first to enter the region. According to the latest information, part of the Vichy regime, which once controlled southern France and collaborated with the Nazis, is holed up in Sigmaringen. The French want to capture the collaborators themselves. All of this seems to put the German nuclear scientists out of Alsos's reach. The Americans therefore initially consider a quick and focused airborne commando raid to round up men like Hahn and Heisenberg. They want to be the first to know what they are still up to, and above all they want to prevent them from somehow falling into Russian hands.

Colonel Pash has plans for a commando action to accomplish this. He has already worked out the plan in detail for a discussion with the General Staff in Reims. Parachutists of the Thirteenth Airborne Division will be charged with landing in the neighborhood of Hechingen and securing the area for at most two Alsos scientists. They will arrive by air and immediately take Heisenberg and Hahn into custody. It stands to reason that Goudsmit will be part of this. For the evacuation of the Alsos men, the soldiers, and the captured scientists, plus documents and materials, a Curtis C-42 transport aircraft will be used. The plan for "Operation Effective" receives approval.

But ultimately, the entire operation is canceled. The Allied land-based forces advance faster than anticipated and the requested aircraft are needed for attacks around Berlin. This does not exactly break Pash's heart, as he later writes, ". . . Frankly, I feared the jump. Still, an Alsos group had to drop with the airborne force. In such a case, I would not call for volunteers unless I was willing to lead them." Goudsmit considers the military nervousness quite exaggerated. The Germans, he repeats to the military staff, have neither an atom bomb nor anything resembling it. Besides, the war is nearing its end. "I considered the German project not worth even the sprained ankle of a single Allied soldier," he comments later.

But Hechingen does remain a priority, and a special task force consisting of infantry and cavalry is set up to get there quickly. On April 20, Adolf Hitler's birthday, the troops leave in the direction of Horb am Neckar, a small town that has already been taken by the French. Beyond that lies the vague no-man's land between the advancing Allies and the retreating Germans. The weather is terrible, it is hailing and snowing, the troops are suffering from the cold, the roads are dangerously slippery, and electricity wires are groaning under a thick layer of frozen drizzle.

Afterward, there is disagreement about the importance of the operation. In his book, Goudsmit refers to the April 1945 undertaking as "Operation Humbug." In his memoirs, written in the 1960s, Pash sticks to the more heroic "Operation Big." Be that as it may, when Pash and his troops finally approach Haigerloch, wary of fierce resistance, the town turns out to be festooned with white pillows hanging from broomsticks and flagpoles, sheets, and other signs of surrender. It is the first German place that directly surrenders to an Alsos unit, Pash notes, but there is not much time for reflection. Haigerloch has been on the Alsos list as one of the most important objectives, and it must be carefully searched at once.

Pash does not have to look for long. At the foot of a cliff that separates the lower from the upper town, there is a bomb-proof concrete entrance to a cave. It is closed with a heavy steel door secured by a large padlock. It is the former beer cellar of the Schwanencafé next door. A sign on the door states the name of the caretaker. When the man is found, he opens the door, but only after Pash has threatened to shoot the lock off and then shoot the man himself. Once in the cave, the Alsos researchers can hardly believe their eyes.

Pash describes the historic moment in his memoirs: "In the main chamber was a concrete pit about ten feet in diameter. Within the pit hung a heavy metal shield covering the top of a thick metal cylinder. The latter contained a pot-shaped vessel, also of heavy metal, about four feet below the floor level. Atop the vessel was a metal frame."

During a first interrogation, the caretaker turns out to be more than willing to talk. This is an experimental uranium pile, he admits, a primitive nuclear reactor with piled-up uranium blocks, a "uranium machine," as the German insists on calling it. Pash and the DuPont chemist Fred Wardenberg (the man who, with Goudsmit, officially concluded that German nuclear research was still in its infancy, in December 1944, in Strasbourg), have a hard time not showing their emotions. Alsos members who subsequently arrive to help with the clearing of the lab ignore all orders and come to have a look at the first German nuclear reactor that has ever been found, there in the Haiger-loch *Bierkeller*. On April 22, the disassembly of the installation begins. Goudsmit is present.

"The scientists had a picnic that day," writes Pash. "Dismantling got under way as soon as the metal cover was removed. Neatly organized in a definite pattern in the center cylinder and caldron were graphite blocks about 18 inches long, 4 inches wide, 2 inches thick. They were placed so as to keep the open center section under the frame. Our men removed the graphite blocks and stacked them for future disposition. . . . At the same time intensive interrogation of the manager and painstaking search was being conducted in an effort to uncover the uranium that had fueled, or was intended to fuel, the pile." Even before that issue had been resolved, Pash had to go on in the direction of Bisingen and Hechingen.

Images of the disassembly may be seen in Sam Goudsmit's enormous photo archive, now housed in the American Institute of Physics in College Park, Maryland, near Washington, DC. Men in combat uniforms and helmets are standing on the structure in the pit. They are handing parts to colleagues who are standing on the edge. No uranium, as it turns out, because none of it is found. In the background, we see steel tanks and a huge metal lid that must have covered the reactor.

The Haigerloch reactor illustrates what the Germans envisage as nuclear reactors at that time. The installation has a concrete external shell about ten feet in diameter, sunk into rock. In it stands an aluminum vat, more than six feet in diameter, and six feet high. Inside that there is a magnesium vat, surrounded by a layer of graphite eighteen inches thick, all together weighing some ten tons. The whole contraption can be covered with a heavy double lid that can be secured with bolts. A total of 664 small blocks of uranium are suspended on chains at regular distances from the bottom of the lid. A quantity of radium can be added at the center in order to supply the neutrons that are supposed to boost the first nuclear reactions.

From the interrogations of German physicists associated with the project, it becomes apparent that the Haigerloch reactor was actually operating in early April. However, the so-called B8 experiment never reached a chain reaction. According to the American experts, the quantities of fissionable uranium and heavy water were too small by a factor of at least one-and-a-half. Whether the reactor did not become critical because of caution or a mathematical error has remained unclear ever since. An isotope analysis in 2015 showed that the uranium used probably came from Czechia and not from the Congolese uranium ore supply taken from Belgium. Moreover, it was not enriched, something that is essential for efficient chain reactions in a reactor or bomb. In short, Haigerloch was almost a caricature of a serious nuclear installation and was, in fact, far below the stature of the physicists who had designed it.

That same afternoon, April 22, Hechingen is captured by French and Moroccan troops. The next day, as Pash moves from Haigerloch to Hechingen with a column consisting of two half-tracks and several jeeps, there is no resistance. The village, full of refugees and homeless people, is calm. At 7:30, the vehicles of the Alsos mission drive into the marketplace, and at once the men begin inspecting laboratories that, according to their information, are located in the town's old woolen mill. They encounter a few well-known physicists: Carl Friedrich von Weizsäcker, Horst Korsching and Erich Bagge, who are later taken to Heidelberg. However, Werner Heisenberg, the presumed leader of the German *Uranverein* (Uranium Club, the name given to the German nuclear weapon project), is not among them.

The Americans do find Heisenberg's office in one of the labs. On his desk is a striking photo of the famous German scientist with four other men. At the far left stands a smiling Sam Goudsmit, dressed in a white summer suit. The photo was taken in the garden of the president of the University of Michigan during the 1939 summer school in Ann Arbor, when Heisenberg stayed with the Goudsmits for several days.

At that moment, Sam is in Alsos headquarters in Heidelberg and oblivious to it all. Afterward he comes in for a lot of ribbing from his companions because of the photo. It looks as if Goudsmit knows the enemy better than any of the other Alsos men. On the same photo, by the way, we see the Italian Enrico Fermi, the Nobelist who emigrated to the United States and somewhat later would lay the basis for the American atom bomb.

In Heidelberg, Goudsmit is coordinating the interrogations of the physicists and technicians from Haigerloch and Hechingen. He will never see the Haigerloch cave in its original state. That same week, zealous American and British officers blow up the entrance of the cave where Heisenberg and his associates tried for the last time to construct a nuclear reactor. Completely senseless, is an angry Sam's opinion when he hears of the destruction. He considers it to be unforgivable to destroy technical evidence that could have confirmed that the Germans had indeed gotten almost nowhere in the nuclear field. Soon afterward it turns out that all the equipment was removed from the cave and the lab before the explosion. In his memoirs, Pash does not say a word about this act of Allied destruction.

In 1947, General Groves returns to the subject in his long list of corrections and comments on the manuscript of Sam's book *Alsos*, which he is allowed to read before its publication. To be safe, Goudsmit sent the manuscript to the Department of War because it may contain classified information. "As advisor I know exactly what I can and cannot say, but I still do not want to end up in jail," he tells his publisher Schuman. According to Groves, there certainly was a reason for blowing up the Haigerloch installations: the French, who controlled the area around Hechingen. "Not to keep them in complete ignorance, but as much as possible," he states in his list of corrections. The sheet of corrections and comments carries one word in big letters: SECRET.

On April 24 the American operation in Hechingen-Bisingen approaches its conclusion, but one place definitely remains on the Alsos list: Tailfingen, where it is reported that Otto Hahn and his team are working. Tailfingen is almost twenty kilometers south of Hechingen, but it is also in an area where it is unclear who is in charge from a military point of view. Pash leaves with a column of vehicles. Now and then he stops in the road, just before the next village center. He phones the mayor. He gives him the choice: surrender or an attack. No one hesitates for a moment. One after another, the villages surrender without resistance. There does seem to be a lack of cooperation from a French unit that is also advancing in the area and blocks the Alsos column for a while. A good conversation, with Pash using his best high-school French, does wonders.

At the beginning of the afternoon of April 24, three hours before the official French occupiers arrive, Pash's column drives into Tailfingen, a small town in a narrow valley. The mayor has already surrendered. In a former school building, the Alsos team comes upon Otto Hahn with his entire team, among them Max von Laue, waiting for what will happen. They are taken to Hechingen for an initial interrogation dealing primarily with the most recent experiments by the group. Where are the reports and other documents about them?

Star player Werner Heisenberg turns out not to be in Hechingen. It is rumored he left on his bicycle days ago to go to his cottage in Urfeld in the Bavarian Alps. His family has already been there all winter, far from a dangerous Berlin.

It is Weizsäcker who breaks first. The desired documents have been carefully packed and sunk down the cesspit of a farm in the neighborhood. Pash and several Alsos researchers go to the spot and use long poles to lift up a tin can so filthy and slippery that they drop it again. In the end, the military personnel of Alsos have to do the job. They take revenge by personally delivering the stinking can to Goudsmit.

In the meantime, it has become known where the uranium for Haigerloch is hidden. According to the German professors, it lies buried in a field not far from Hechingen, very near a grist mill. In a freshly plowed field, the Americans soon find three drums of heavy water, and somewhat farther on a big wooden chest filled with blocks of metallic uranium.

Pash writes: "The uranium metal was so heavy that the men formed a human chain to pass the ingots one by one to a spot from which they could be loaded on trucks. When they were stacked ready for loading, it was hard to realize that the small pile, measuring not more than two cubic feet, would weigh about two tons." The uranium, heavy water, and graphite that have been found are taken to headquarters for transport to the United States. There the German uranium is ultimately used in the Americans' own nuclear program.

"Alsos has hit the jackpot," Colonel Pash signals to headquarters that day. Supreme Commander of SHAEF Eisenhower adopts the words literally. "Boris Pash has hit the jackpot," he reports to the president in Washington with a sigh of relief.

In Heidelberg, Sam concentrates on the men he has found in Hechingen and its surroundings. The group is a veritable representative group of German physics, which Goudsmit knows like no other: old Otto Hahn is there, the diplomat's son and physicist Carl von Weizsäcker, nuclear physicist Karl Wirtz, and Nobelist Max von Laue. Two younger scientists also seem important, because they have developed methods to separate isotopes: Ernst Bagge and Horst Korsching. Six German physicists in total are taken into custody and taken to Heidelberg, which is in American hands. There is no real question of French interference. Insofar as the French are active in the area, Goudsmit later says scornfully, they are more interested in the wine cellars, the chickens, and the pigs than in nuclear secrets and scientists. There are no Russians to be seen anywhere.

That leaves just Heisenberg. The village of Urfeld, where it is rumored he is staying, is in territory still held by the Germans, a bit south of the Bavarian capital of Munich, but at this time, in fact, only that city is being defended. Colonel Pash decides to take a risk, going with six men. He drives from Hechingen through the foothills to Urfeld. On May 1, they reach the village without incident and soon find the Heisenberg cottage. The physicist is standing on the porch and welcomes the gentlemen. He invites them inside, where his wife and two young children shyly shake hands with the Americans. "At that moment, I took a deep breath," Pash writes later. "Alsos was about to close the book on one of the most successful intelligence missions of the war—or so I thought."

He informs Heisenberg that he has to come along to Heidelberg to be interrogated about his work during the war. The German scientist agrees at once, especially when he hears whom he will be speaking with: *Herr Kollege* Goudsmit. He gathers together several file folders with calculations and notes made in Haigerloch and places them before Pash on the kitchen table. Pash puts them in his briefcase.

Almost immediately after this, a few German officers appear in Urfeld, wishing to surrender to Pash with their units, not realizing there are only seven Americans, who would be easy to overpower. On the spot, Pash thinks up a ruse: he proposes that the surrender be arranged formally the next day and sends the German officers on their way. As soon as they are out of sight, the Americans quietly leave the village.

That evening, Pash's caution turns out not to have been exaggerated. Around eleven o'clock about a hundred SS men carry out an attack on Urfeld, in search of the Americans. When these turn out to have flown the coop, the SS men take out their anger on the home guards for failing to resist the Americans; there are several deaths. The next day the corpses are floating in the lake beside the pretty Alpine village.

Two days later, on May 3, Urfeld is officially occupied by the US Army, and Pash and his men call on Heisenberg again. He is ready to accompany them, but he makes a request that initially sounds strange. He would prefer to be taken into custody with a show of force, so as to keep his fellow villagers from getting the impression that he is going along voluntarily. In this way, he hopes, his wife and children will have the least possible trouble with the neighbors.

Pash thinks this is nonsense and speaks firmly to Heisenberg about it. The German-speaking neighbors, he writes with some irony in his memoirs, can easily have understood the English conversation as rather unfriendly. And it was meant that way.

Later that day in Urfeld, 700 Germans surrender to the Americans. Elsewhere in the village, soldiers find the corpse of Colin Ross in a house belonging to Nazi youth leader Baldur von Schirach. An Austrian author and filmmaker of Scottish descent who became a prominent Nazi propagandist, Ross and his wife had committed suicide. Three days before the grisly discovery, on April 30, the Führer had done the same in Berlin.

Also on May 3, 1945, Munich falls. Two men who are also on Goudsmit's nuclear list are picked up in that city. To begin with there is Walther Gerlach, head of German nuclear research. With him is Kurt Diebner, the man who built the previously located Stadtilm reactor. The last missing tons of uranium from Stadtilm are also found in Munich. The German uranium is transported to the United States, where it will ultimately be used in the American atom-bomb program.

Later that week, Goudsmit and his staff are on the front steps of Alsos headquarters in Heidelberg when Heisenberg arrives in a convoy from Hechingen. The first meeting between the two is unreal, after all those war years without contact. Sam knows Heisenberg well, from Leiden and from the summer schools later in Ann Arbor. The friendship is still there, and there is also mutual relief that both have survived the war. And yet everything has changed. "It was sad and ironic listening to him . . . , when I was aware how much more we knew about the problem than he did," Sam writes in *Alsos*.

Heisenberg still does not know this. He offers to inform the Americans about nuclear fission and to show them the results of the research in his laboratory. There is a brief discussion of his having to go to America, after everything that has happened. But he declines. "Germany needs me," he says firmly.

It is the same blend of self-importance and nationalism that will later regularly enrage Sam Goudsmit. To be sure, Heisenberg is no Nazi. Up to a point, he has even been a critic of the regime. At the same time, he was a critic not on principled grounds but chiefly to pilot German science through the war with as little damage as possible. Afterward, he is convinced, everything will return to normal once again.

In the years after the war, Goudsmit and Heisenberg will regularly discuss why the German physicists accomplished so little in the nuclear field. According to Goudsmit, this was the result of ignorance and the fact that a totalitarian regime does not offer space to a free science. According to Heisenberg, it was much more a conscious strategy on his part, and that of other physicists, to make as little progress as possible toward the development of an atom bomb. The two men will never agree. In any case, the stacks of nuclear documents that Alsos collected in Germany do not exactly conjure up the image of a self-restrained community of physicists.

Heisenberg and the other physicists from Hechingen, Tailfingen, and Munich are confined separately in the rooms of a confiscated villa on Philosophenweg in Heidelberg, where they are guarded by black American GIs. This leads to occasional complaints from the prisoners, Goudsmit says later with a mixture of irritation and amusement: "They had perhaps lived too long under the myth of Aryanism."

The physicist Walther Gerlach, captured in Munich, seems to be greatly relieved now that he is in the hands of the enemy. Goudsmit knows this German well, too. He met him a number of times before the war, in England among other places. In those days, Gerlach was always rather guarded, particularly because, as a committed nationalist, he regarded German science as all-powerful.

It is Gerlach who, in the late stages of the war, directed the attention of Nazi minister and chief architect Albert Speer to the necessity of finally taking nuclear research seriously, and then got the green light for a catch-up effort under his leadership. Gerlach is a born leader, authoritarian but beloved for his quick and sharp insight. No Nazi, but just like Heisenberg concerned, above all, for the continued existence of German science. He made a series of wrong choices during the war, Goudsmit later judges mildly.

Goudsmit finds that Gerlach has aged a lot since he last saw him. But he talks openly and with relief about the German nuclear efforts of the last few years, and also about the fact that he and Diebner have had to struggle against the arrogant clique of theoreticians around Nobelist Heisenberg. Diebner himself, taken to Heidelberg along with Gerlach, is his complete opposite. He is somber and hostile, not prepared to say much, not even to a fellow physicist like Sam Goudsmit. "He was as sullen as a real prisoner," Goudsmit concludes.

During the months that follow, Diebner, now confined with nine other German physicists in a country house in England that is full of hidden British microphones, suffers from full-fledged depression. The cause is the news on August 6 that the Americans have destroyed the Japanese city of Hiroshima with a nuclear weapon. In the aftermath of that event, a discussion among the Germans is recorded. The resulting "Farm Hall tapes" will be released only in the 1990s. The recordings once again confirm the image of a group of outstanding scientists who, assuming they had all the time in the world, let nuclear fission completely slip through their fingers after discovering it in 1939. After all, they seem to have reasoned, American science did not amount to much.

In July 1945, two months after Hitler's suicide and the German capitulation, Goudsmit and his men, after long insistence, get permission from the Russians to visit Berlin. The center of the city, the locus of Nazi power, is a ruin. In it, the Allies watch each other's movements closely. The suburb of Dahlem, with its countless scientific institutes, has remained largely undamaged. Sam visits the Kaiser Wilhelm Institute for Physics, where Einstein once lived and worked. Professor Fritz Haber, who developed poison gas during the First World War, worked in the building next door.

Somewhat farther on is the Institute for Physical Chemistry. It was there that, in 1939, Otto Hahn, with Otto Frisch and Lise Meitner, was the first to demonstrate the fission of uranium nuclei and the release of neutrons, which constitutes the basis for an energetic chain reaction. Hahn's institute, a tall villa with pointed steeples on the corners, has been hit during a bombing raid and stands there scorched and empty. Many of the windows have been nailed shut.

The head of the information service of the American occupation forces in Berlin has moved into the undamaged physics institute. It is a large, block-shaped building with neo-classical white ridge-pieces and columns. When the Americans moved in, an officer at the entrance says, the Russians who had conquered the city had already emptied the premises. A lot of scrap metal has been left behind in the building's backyard. Among it are a number of measuring instruments for use in nuclear research. First-rate equipment, Sam notes. Here and here, mixed up in the junk, he even finds briquettes of compressed uranium oxide, evidently left over from experiments that were carried out in the bunker cellar of the building. According to the officer, there is a remarkable round swimming pool, partly covered with wood planking, in the cellar. He has no idea exactly what it is.

That mysterious swimming pool is located at the foot of a concrete staircase near the side of the lab. It can be only one thing: a pit in which a nuclear reactor stood or should have stood. Everything is still there. The metal containers, crane, and frames from which the blocks of uranium could have been hung. When all of this was built and by whom, and how far they had gotten, is unclear. Probably it was Diebner, who had fled Berlin in 1943 with his material and staff because of the growing number of bombing raids. He was never again able to carry out serious experiments elsewhere in an increasingly battered Germany. Evidently the Berlin reactor was not yet operating when it had to be taken apart and moved south.

That July day in 1945, Goudsmit sends everybody else away and in the quiet twilight realizes as never before how differently history might have proceeded. If Paul Harteck and Wilhelm Groth had found a hearing from the authorities with their 1941 letter about atomic explosions. If Diebner had managed to get the Berlin reactor operating, or the second version of it in Stadtilm. If Gerlach had been able to win the armaments minister over to his side before 1944. If Heisenberg had been less of an arrogant theoretician, and Haigerloch more a center of real experimentation. Then the German atom bomb would surely have been a fact.

With that thought, a shiver runs down his spine.

More than ten years later, on Saturday October 11, 1958, members of the Alsos mission have a reunion in Washington's Shoreham Hotel, mainly upon the initiative of their former leader in Europe, retired Colonel Boris Pash. A few weeks later, Sam writes him a note of thanks. "Too bad that there were not more civilian Alsos people there, but they are undoubtedly in the same boat as I am: broke or almost so."

Some of those he missed at the reunion were his right-hand man Fred Wardenberg of Dupont and his science advisor Wallace Brody, both chemists. Maybe they will show up the next time, he says, and he is happy that the plan is to get together again two years hence. "In any case, that is better than in three years." In the meantime, a small publication will be prepared in order to maintain contact.

The most notable absentee is Lieutenant-General Leslie Groves, the military commander of both the Manhattan Project and the Alsos mission, which in fact was named after him. In a short letter, Groves thanks Pash for the invitation, which he has to decline because of a trip to Madrid. "How ironic it is that this time I am in Europe and you in Washington. Whereby I have the advantage that this time nobody in Washington is constantly asking for advice or giving orders."

Present in Washington, with his wife and several colleagues, is Sam's permanent information officer, Robert R. Furman. They met for the first time in London in 1944. Furman did not leave his side. After the war, the "Mysterious Major," as Sam consistently continues to call him, headed a successful construction firm in Maryland without ever talking about his war years.

In the run-up to the reunion, Sam helped Colonel Pash with drawing up the list of invited guests. During that time, the colonel more than once drew upon Sam's famous address books. A letter in the archives concludes one of those lists with a striking joke: "Dear Boris. Should we not invite the colonel whose car I stole during the landing at Cherbourg? I have forgotten his name, but I think it was Colonel Ford, who later served with G-2 [Military Intelligence] in the 5th Army in Chicago." Sam is referring to the military truck in which he himself, formally a civilian, rode to Paris in August 1944.

At the beginning of September 1958, as the invitations are going out the door, a supplementary plan takes shape. Colonel Pash writes to a general in the White House with whom he is on friendly terms that it would be really nice if, on the occasion of their first reunion since the war, the former members of the Alsos mission could meet the incumbent president on October 11. To be on the safe side, Pash sketches the history of Alsos once more, and the mission's nuclear detective work in Europe. "The unit was formed on November 23, 1943, and disbanded on November 15, 1945. The unit numbered 201 members in total, of whom 55 American scientists, 12 British scientists, and 4 Allied scientists; 55 officers, of whom 33 technical specialists, of whom 10 US Navy; 22 counter-espionage experts; 51 enlisted men; and 4 administrative personnel." He states that he expects at least forty people at the reunion, among them Wallace Brody and Sam Goudsmit.

"I would like to propose the visit to the President because the men of Alsos and their unit have never received recognition of their excellent work during the Second World War. A meeting will be a reward that will honor the unit and which the men will always remember."

The meeting with President and General Dwight Eisenhower, who was the commander-in-chief, SHAEF, during the Alsos operation, did not take place that Saturday in October. The evening before, the forty Alsos men enjoyed cocktails with the colonel and his wife at the Pash home in McLean, Virginia. In their meeting the following morning, they adopt a formal resolution. It urges the official recognition of the special insignia that the members of Alsos wore on their shoulders during the war: the Lightning *A*, a white Greek letter alpha on a light blue square background, with a red lightning bolt through the alpha.

In January 1959, the Department of Defense responds with a brief refusal. There seems to be no trace of the insignia in the heraldic archives of the US Army, not even of its unofficial use. That is a problem. A formal recognition of the insignia would mean its approval, which means that an official design with drawings and specifications has to come from a manufacturer. "This is a request that has not been made before."

On the other hand, the note of refusal continues in a gallant spirit, there is great appreciation of the efforts to secure more recognition for Alsos. "For that reason, there is no objection at all to using the shoulder insignia in connection with the present-day activities of the former members of the unit.'"

CHAPTER 6

THE TRUTH OF KISTEMAKER

Amsterdam, 1960–1961

O n Friday afternoon, October 14, 1960, the Dutch Communist daily newspaper *De Waarheid* kicks off what turns out to be one of the most high-profile controversies in the post-war Netherlands. "Amsterdam laboratory cooperates in the preparation of German A-bomb," is the headline above the article, which continues on page 4. The account simultaneously touches all the exposed nerves inside and outside the Communist movement. Fifteen years after the war, the development of a new German army is still not widely accepted. The newspaper still has daily accounts of the convictions of war criminals. A German atom bomb is too much. Moreover, the nuclear weapons under discussion will, of course, ultimately be aimed at the Soviet Union and the Communist brothers and sisters in East Germany. The Dutch Communists are furious.

Other stories in the same newspaper report massive worker demonstrations in Tokyo, and a speech by Soviet leader Nikita Khrushchev, made in the United Nations on October 12, in support of a resolution to extend independence to colonial countries, which was greeted with loud applause. "Even the Spanish Franco delegation had to clap their hands," according to the newspaper's correspondent at the UN meeting in New York. The resolution, *De Waarheid* claims, is a serious defeat for the Americans, although the US representative to the UN, James Wadsworth, will eventually claim to have applauded.

The article about the Amsterdam laboratory is written by "a reporter," who later turns out to be the journalist Wim Klinkenberg. He reveals that the Laboratory for Mass-Spectrography "under the leadership of Prof. Dr. Kistemaker" has already been actively cooperating for two years in the preparation of nuclear weapons for a new *Wehrmacht* in Germany.

The run-up to the news story in *De Waarheid* begins on Wednesday, October 12, with reports in the American press about West Germany's efforts to produce nuclear weapons. The Germans, the story goes, are working on a purification technology for uranium. In this way, the type of uranium that can be used in a nuclear reactor, uranium-235, can be separated from natural uranium in ultrafast centrifuges.

In the meantime, US President Eisenhower has confirmed the reports and hinted that he has no objection to the development, even though it contravenes agreements made soon after the war to allow Germany no military capability.

In the lab on the eastern outskirts of Amsterdam, property of the Municipal University of Amsterdam, work is taking place on the same technology to purify uranium on which the Germans are said to be working, the so-called ultracentrifuge. Jacob Kistemaker heads the new lab, which has been located on Kruislaan since 1959. Kistemaker, born in 1917, is a striking, taciturn man, a farmer's son from Kolhorn, near Den Helder, who had a talent for mathematics and studied physics in Leiden. His laboratory is located in an open polder and surrounded by a high fence.

For years he has worked on so-called isotope separation. Atoms of the same chemical element can have a somewhat different mass and thus different nuclear characteristics. In the case of uranium, uranium-235 is easily fissionable and uranium-238 is not or barely so. Therefore, uranium-235 is essential for achieving a nuclear reaction, but only a low percentage of uranium-235 is found in natural uranium.

Initially, Kistemaker experimented with potential isotope separation technologies in a hall on Hoogte Kadijk belonging to the Municipal Electricity Company in Amsterdam. The link with the electricity provider was convenient; Kistemaker's tests require very high tension levels. The objective is the enrichment of natural uranium, in which the quantity of uranium-235 nuclei is increased. Minimally, a 3 percent enrichment of the uranium-235 content is necessary for a nuclear chain reaction. In the lab on Hoogte Kadijk, Kistemaker, then employed by the physics foundation Fundamenteel Onderzoek der Materie (FOM), built the first mass-spectrograph with the assistance of a number of initiates. This is a huge electrical magnet that forces a bundle of atoms into a curve. In this process, a heavy atom is bent less easily than a lighter atom. The differently weighted atoms therefore land in different spots on a target at the end of the installation, the light ones on the outside edge of the curve, the heavier ones closer to the center.

The Dutch experiment in an old factory hall suddenly became international news in 1953. Kistemaker presented the first ten milligrams radioactive and fissionable uranium-235 produced outside the United States. Scientific greats such as Irène Joliot-Curie (daughter of Madame Curie), Niels Bohr, and Ernest Lawrence, all three of them Nobelists, visited the lab for the sole purpose of seeing Kistemaker's feat with their own eyes. However minimal the quantity of enriched uranium may have been, the Amsterdam success is regarded by historians as a key reason for the United States to release their isotope-separation technology in the 1950s: keeping it secret made no sense anymore. At the end of 1953, President Eisenhower delivered his famous "atoms for peace" address to the United Nations. The atom, he said, must change from being a feared weapon to a harbinger of welfare and prosperity. The availability of lightly enriched uranium, begun by Kistemaker's demonstration, ushered in the beginning of the civilian nuclear industry.

The opinion of the *De Waarheid* reporter is simple: according to post-war agreements, the Germans are not allowed to manufacture nuclear weapons. But the government led by Konrad Adenauer is talking openly about industrial applications of nuclear energy. Moreover, the "German revanchists" are not banned from bringing in nuclear weapons from outside, the thirty-seven-year-old Klinkenberg writes in the customary *Waarheid* blend of reportage and commentary. The contacts of Kistemaker and his German colleagues are presumed to be crucial in industrial applications.

Ever since the American atom bombs fell on Hiroshima and Nagasaki, Kistemaker had been intrigued by the idea of enriching uranium. When still a graduate student, he was one of the first Dutch scientists to read the "Smyth Report," about the development of the American atom bombs with the Manhattan Project. Dating from 1944, that report was intended for internal use, but soon after the war several copies were already circulating within a restricted circle of scientists. Kistemaker was one of the few to realize how essential the industrial enrichment of uranium had been for America's success in this undertaking.

Besides, he recognized a technology he first heard about in the early 1950s during a visit to a still-damaged Germany. It was in the Hamburg physics lab of Paul Harteck and Wilhelm Groth, two names that also showed up in Sam Goudsmit's Alsos notes. In 1939, Groth was one of the first to tell the German Nazi authorities that recently discovered nuclear fission had the potential of yielding an extremely powerful explosive.

Kistemaker, who died in 2010, insisted into old age that the meeting in Hamburg was really a coincidence, and in any case, informal: as a curious young physicist he walked into the lab and, unnoticed, joined a colloquium where the necessary centrifuge technology was being discussed. "The door happened to be open," he insisted in interviews, invariably dressed in his white lab coat and seated behind his desk in the FOM Institute.

Once he was back home, Kistemaker radically changed the course of his isotope research. He realized that separating isotopes with mass spectrographs will never yield enough material. The future belongs to the centrifuge. The mechanical stability of the installations is crucial in this. The centrifuges not only have to remain intact at speeds of thousands of revolutions per second, but they also have to revolve without a tremor in their bearings. Because of the concentrated spinning energy, every deviation is fatal. Therefore, part of the work on Kruislaan consisted of a search for the correct measurements and the design of bearings that will be proof against every force. Ultimately, it was Kistemaker and his team who took the first steps toward the modern ultracentrifuge, and thereby laid the basis for Urenco in Almelo, a Dutch-German-British facility for the enrichment of uranium. The factory works for the civilian sector and furnishes material for fuel rods in nuclear power plants. In the halls in Almelo, there are thousands of slender, linked centrifuges on magnetic bearings that separate uranium according to weight. In the late 1970s, the plant got into the news when it became evident that the Pakistani spy Abdul Khan, later the head of nuclear research in that country, was able to walk out the door with a briefcase full of secrets relating to the ultracentrifuge. It is suspected that this knowledge not only made the Pakistani atom-bomb program possible, but later also helped Iran get hold of the technological know-how.

In 1960, *De Waarheid* is still worried about a completely different perspective on Kistemaker's work on the ultracentrifuge. That he is working on it is one thing, but it is mainly his choice of scientific friends that arouses amazement at the newspaper and in other circles. It is whispered that he is working not only with the Americans but also with colleagues and the authorities in Germany. When the news about the German ultracentrifuges breaks, a bomb goes off in Amsterdam.

Exactly how far the cooperation of the FOM institute with

the Germans extends, reporter Klinkenberg does not really know at that moment. He has suspicions, however, based above all on a number of visits of laboratory staff to the University of Bonn and the Degussa firm in Düsseldorf. In 1959, too, the German minister of nuclear affairs, Siegfried Balke, visited the lab in Amsterdam. Groth, a German academic, once came by with thirty students, *De Waarheid* writes. In an interview not long before his death, Kistemaker himself vigorously rejected charges that he helped the Germans with the development of uranium enrichment. When he spoke with Degussa, for example, it was about magnets for electric motors, he says. "They were good at that."

However, at the outset of the 1960s, he definitely acts ineptly when *De Waarheid* opens a full-throated attack on him. When Klinkenberg phones him a few days before publication about the rumors of German visits to his Kruislaan lab, a testy Kistemaker denies everything. In an interview with *De Telegraaf*, the journalistic and political polar opposite of *De Waarheid*, he does say that he considers Dutch-German cooperation in connection with the ultracentrifuge to be "quite desirable." This remark pours oil on the flames, the more so because the authoritarian Kistemaker not only has friends in the lab but also enemies. The turbulent 1960s are at hand; increasingly, students and progressive university staff are challenging the old academic relationships, on Kruislaan as well.

In the week after the news stories about the German connections to the nuclear lab in Amsterdam, Klinkenberg gets a tip about an issue that could really embarrass Kistemaker. It remains unknown who was the source of the tip. On Tuesday, October 18, *De Waarheid* again has the Amsterdam nuclear research and the Germans on its front page. But this time it is Kistemaker himself who has to take the heat. In the post-war Netherlands, the headline could hardly be more devastating: "During the war Kistemaker worked in Paris for the Germans."

A subhead deliberately reinforces the effect: "Now cooperating with Hitler's A-bomb advisors."

The story deals with Kistemaker's visits to occupied Paris from 1942 to 1944. As a graduate student in physics at Leiden University, working in the Kamerlingh Onnes Laboratory, Kistemaker was stationed in Paris as a scientific staff member of Cellastic, a secret German organization. According to *De Waarheid*, this was under the direction of the naturalized Dutch Nazi H. W. F. Kleiter. Cellastic's most important assignment was the collection of information about French nuclear research, which was under the supervision of Professor Frédéric Joliot-Curie. He, "as is well-known," knew how to keep the Germans at a distance, *De Waarheid* immediately adds. Its target is Kistemaker, who, the newspaper charges, worked for Cellastic as a young man in his mid-twenties until the liberation of Paris in August 1944. "With this, his current activity as leader of ultracentrifuge research appears in a much more lurid light."

Jaap Kistemaker received his Leiden doctorate in 1945 with a dissertation on the heat qualities of the element helium. He left for Denmark, where he worked for a year with the famous theoretician Niels Bohr in Copenhagen. In 1947, he entered the employment of the FOM physics institute. Among other tasks, he made a professional trip to America, where he visited the nuclear lab in Oak Ridge. In 1955, he was promoted to the rank of professor at Leiden; he also became director of the new FOM lab for mass spectroscopy in Amsterdam. There the successful work on isotope separation that began in the temporary lab on Hoogte Kadijk was continued in a real lab. At that point Kistemaker was thirty-eight years old.

What almost nobody knew in those years is that Kistemaker was called on the carpet by Dutch Military Intelligence soon after the war to justify a three-months' stay in Paris early in 1944. Sam Goudsmit attended a number of those interrogations, and that was no coincidence. After the liberation of the French capital in 1944, Sam and his Alsos team discovered that at least four Dutchmen had worked at the German company Cellastic at 20 Rue Quentin-Bauchart, a quiet side street south of the Avenue des Champs Élysées. One of those Dutchmen was Kistemaker, and the question is what exactly was he doing there. For example, was he aware of the secret intentions the Germans had for Cellastic? And did that have anything to do with nuclear research?

During the post-war interrogations in The Hague, the twenty-eight-year old Kistemaker admitted he did realize the Germans' intentions for Cellastic. But he made it appear that he was in fact practicing counterespionage by spilling the beans back in Leiden about the real intentions of Cellastic. At the same time, he said, the work at Cellastic was actually pretty boring and not very challenging. The high point for him was a meeting with the great Joliot-Curie in 1943. (He, in turn, was little impressed by his young Dutch visitor and wondered what Kistemaker was up to.) An excellent salary of several hundred guilders per month, and the stay in Paris, compensated for the boring work. Thanks to administrative errors, he sometimes got paid even more, without anyone ever asking that the excess be repaid. Cellastic was a piece of luck; an adventure abroad, furnished with all the necessary documents, a bonus in the gray and anxious war years.

In December 1960, the minister of education, Jozef Cals, answers questions in the Second Chamber concerning Kistemaker from members Cees Hazenbosch (Anti-Revolutionary Party) and Marcus Bakker (Communist Party). Kistemaker, Cals states, did not commit blameworthy actions during the war. The graduate student had taken employment with Cellastic "for scientific purposes" on the request of the director of the Kamerlingh Onnes Laboratory.

It does indeed appear that Kistemaker was invited to make his Parisian excursion by the Leiden energy physicist Jan Ketelaar. Ketelaar was the first of the Leiden physicists to respond to a Cellastic advertisement for scientific experts and got a contract. He, too, was in Paris for a few weeks on several occasions, was paid for this, and provided his assistant Jan Willem Zwartsenberg with a similarly lucrative and relatively easy job. Ketelaar, later professor in Utrecht and attached to the KEMA-lab of the electricity companies in Arnhem, died in 2001. The obituary that appears in the publication of the Academy of Sciences dodges the question of his part in the Cellastic affair. It is noted that Ketelaar used his trips partly to get into touch with refugees and to smuggle medical drugs.

After the interrogations, the conclusion was that Kistemaker did work for the dubious organization, but that he did so out of naiveté rather than anything else, more out of greed and a youthful wish for adventure than out of genuinely criminal motives. There is even some praise for the way in which, in the middle of the war, Kistemaker was able to talk his way to securing, in Paris, a quantity of helium for his lab in Leiden. Former colleagues from Leiden said that they did not remember Kistemaker as a collaborator but rather as a level-headed pragmatist and a hard worker. And this in a lab where work continued during the Occupation as best it could, under the leadership of Wander de Haas. It is later determined that Kistemaker did advise his students to sign the 1943 loyalty declaration, which committed them to "abstain from all actions against the German Reich." Male students who did not sign (a large majority) were subject to forced labor in Germany; many of them went into hiding.

After all this, the Cellastic affair played no part in Kistemaker's appointments in Leiden and Amsterdam; the dossier disappeared into a desk drawer. FOM claimed later not to have been aware of the investigation into his war years.

The head of the Kamerlingh Onnes Laboratory in the war years, Professor Wander de Haas, accepted three invitations to go to Paris, accompanied by his wife, and allowed himself to be well paid for his services. During his third trip, however, De Haas and his wife quietly slipped away to Switzerland and then went on to London. There he reported to the government-in-exile but neglected to mention the existence of Cellastic. He did boast about the concealment of a quantity of uranium ore (so-called "yellowcake") in the lab in Leiden, organized via a Belgian connection. After the war, he was reproached for showing extreme stupidity. To his great outrage, he was deprived of his title of professor for a year.

When the Cellastic affair explodes in the Netherlands in 1960, fifteen years after the end of the war, Sam Goudsmit has recently resigned his position as head of the Physics Department in Brookhaven National Laboratory, Long Island. His active research has increasingly given way to his busy schedule as editor-in-chief of the ever more important *Physical Review*, published by the American Physical Society. The years are beginning to take their toll, even for the always-energetic Sam Goudsmit, then fifty-eight years old. As editor-in-chief, he still reads virtually all manuscripts that come in. The other work he is glad to leave to his co-workers, but he does want to be informed about things. The information that his old compatriot Jacob Kistemaker in Amsterdam has gotten into trouble because of Communists working at the daily *De Waarheid* escapes him for the first few weeks.

But when Kistemaker's wartime history also begins to play a role, Goudsmit is drawn willy-nilly into the matter. Dutch journalists discover fairly soon that in Sam's 1947 book about the Alsos mission there is a chapter about a small group of Leiden scientists who worked for a shady technical business in Paris. According to Goudsmit, this firm, Cellastic, was a cover for German industrial espionage and wartime theft, specifically in France. Officially, it gathered and traded patent information. In reality, it established contacts with industrial and scientific labs and explored what might be happening there that could be of importance for German industry and the war effort.

One of those Dutchmen was a young doctoral student from Leiden, who is referred to as Dr. K., but who is unmistakably Jacob Kistemaker. "Young and green," Goudsmit calls him. In *Alsos*, the chapter about Cellastic occupies a mere sixteen pages, which are a digression, moreover, from the story Goudsmit really wants to tell: why German scientists failed in the development of their own atom bomb, even though they had themselves discovered nuclear fission in 1939.

When Sam entered Paris via the Porte d'Orléans in the late summer of 1944, two days after the liberation of the city, he looked up Professor Frédéric Joliot-Curie as soon as possible. It is already known that he quietly played a significant role in the French resistance without really dirtying his hands. During Goudsmit's visit, his laboratory at the Collège de France turned out to be full of explosives. It had been used to make Molotov cocktails.

During the years just past, Joliet-Curie tried to evade as much as possible any German interest in French nuclear research. In Paris, Joliet-Curie was using one of the earliest cyclotrons, a magnetic circuit to accelerate particles. Under the supervision of Erich Schumann and Kurt Diebner, the Germans considered moving the entire apparatus to Berlin, but ultimately left it in Paris, where German physicists sometimes used it for experiments. The presence of Germans in his laboratory earned Joliot-Curie, however unfairly, the name of collaborator in those days. Sam and his staff examined the cyclotron lab for indications of atom-bomb work. They found a few letters and notes that had been left behind. These turned out to be either meaningless or private correspondence in which the sender asks her loved one whether he can bring along a bottle of Chanel No. 5 when he returns to Germany.

Paris itself was an unalloyed disappointment for the Alsos mission. There were no serious indications of nuclear interest among the Germans. And advancing was not yet possible; farther north and east, the war was still raging. In order to escape boredom, the Americans began to search through the offices of the German industrial firms in the city. Who knows: something might turn up.

And so it did, during those first few days in Paris. In the piles of paper that came in for examination was an invoice for parts with a connection to radar equipment. Goudsmit tracked down the engineer who signed the invoice and questioned him in congenial surroundings, over dinner with lots of wine. The man, who had a scientific education, understood quickly that he was dealing with professional colleagues and began to talk. Just as Goudsmit was carrying on a scientific intelligence mission, the Germans were doing something similar during the war. The engineer worked for them himself, although working was a big word, he said: now and then he informed them about French inventions and inventors. The organization was called Cellastic and had its office at 20 Rue Quentin-Bauchart. It seemed to be a patent office, the Frenchman says, but it was actually an espionage service. Moreover, not only Frenchmen worked there, but also a few Dutchmen and Swiss.

Goudsmit paid a visit to the address he was given. It turned out to be a small office in the former Venezuelan embassy in Paris. It was conspicuously empty, but it was not an ordinary office. The rooms had been soundproofed. There were internal telephones of a kind that cannot be overheard. And upstairs there were vestiges of a simple chemical laboratory.

Goudsmit's detective instinct woke up. In the run-up to the Alsos mission, he was specially trained in America for intelligence work and technical investigation. The tips and tricks supplied to him now stood him in good stead. Next to the switchboard he found a map of the rooms and a list of the occupants of the desks, with their expertise indicated. There also turned out to be a record of all the incoming calls of the two most recent months. The library pointed to a strong interest in science and technology. There was unopened mail in the mailbox, delivered after the Cellastic staff left. And he found a piece of carbon paper that was used to type the names and addresses of the employees.

The porter had a Cellastic visitors list for the last couple of weeks, with names, dates and times. Dutch entries are found on an appointment calendar, appointments for visits to French scientists at the Sorbonne. On this calendar, Goudsmit found the names of two Dutch scientists whom he knew personally from his Leiden years. He deliberately did not mention those names in his book. As became known later, they were the names of Wander de Haas, director of the Kamerlingh Onnes Laboratory in Leiden, and Jan Ketelaar, a physicist in that lab. In the list of staff members, there were also references to two young Dutchmen: Jan Willem Zwartsenberg and Jacob Kistemaker.

Whether the Leiden physicists were still in Paris at that point in 1944 was unclear. Goudsmit visited a scientific book store in the neighborhood and asked whether any Dutchmen had ever been noticed there. Sure, the book dealer said. But a parcel he recently sent to Monsieur Zwartsenberg unfortunately came back as undeliverable. He did get a letter about the parcel from the Dutchman a few days ago, sent from a village near Rouen in Normandy. When Goudsmit rang the doorbell of the house in Pont-Saint-Pierre the next day, the door was opened by a young man whom "I recognized at once as a Dutch student." Initially "Zwart" refused to say anything much, but when Goudsmit unexpectedly launched into Dutch to denounce the young physicist ("a ... collaborator, a quisling, a man from my own university, the University of Leyden") Zwartsenberg broke down. He confessed that he had the time of his life in Paris. He was well paid, had a car, little serious work to do, and above all, lots of free time. He agreed that it was indeed improper that he, a student, called himself "Dr." on his card and claimed membership in an organic chemistry laboratory.

It was all the more painful that this was the lab of Professor Anton Eduard van Arkel. The pharmacologist Van Arkel was a leading figure in the Dutch resistance. When Kistemaker, as one of the first post-war graduate students to get his Leiden doctorate, defended his dissertation, it was none other than Van Arkel who gave him a rough ride. Although that was thoroughly normal on such occasions, Kistemaker would complain about it for years afterward in letters to friends and colleagues, because he suspected that the wartime affair played a part in it.

After Zwart's confessions, Sam felt depressed: "Meeting a traitor, even a mild one, is so much worse and so much more incomprehensible than meeting an enemy." Back in Paris, he informed the Dutch military attaché about all the Dutch employees of Cellastic.

The name Cellastic showed up for the first time in Paris, but it had a Dutch background. The Cellastic-Holland office was founded in 1939 as a subsidiary of the firm Amsterdam Goederenhandel, owned by J. Abs, a German banker from Bonn. Abs came to the Netherlands for the Deutsche Bank in 1924. Before that, he was the attorney for the NV Handelsmaatschappij Cellastic, with headquarters at 26 Hertoginnelaan in The Hague. During the war, a naturalized German with pronounced Nazi sympathies was in charge: Hans Kleiter. He also established the Paris office but left its daily operation to an assistant.

In the aftermath of the war, the office of Cellastic-Holland was turned inside out by the military intelligence services. Sam Goudsmit was present at the interrogation of Kistemaker, based on the Cellastic material, in the autumn of 1945. In the report about the affair, the ultimate assessment was fairly mild. The conclusion was that Kistemaker was naive rather than a collaborator. No measures followed, and the affair was forgotten. Until October 1960, that is.

When Goudsmit was in the Netherlands in the autumn of 1945 for the interrogations of Kistemaker and De Haas, he stayed in The Hague's Hotel des Indes. That was not only just around the corner from the house on Prinsestraat where he was born but also where his grandfather worked as a tourist guide. (The story of the humble origins of the Goudsmits invariably comes up in his memoirs and recollections.) At least as remarkable was that the physicist Jan Ketelaar searched him out in his hotel room with a request for help. Ketelaar was accused of working for German-owned Cellastic in Paris and hoped that Goudsmit would exonerate him. Evidently, he did not realize how well Goudsmit knew the affair. And Sam reacted firmly. He was unable to help Ketelaar for the simple reason that he did not have the full details of his file and also because he lacked the authority or contacts to arrange anything. "I never heard from him again."

At the end of 1960, the Kistemaker affair gets further out of hand. Initially, *De Waarheid* is the only newspaper that, time and again, keeps fanning the affair in the aftermath of American news that West Germany may want to work on a nuclear weapon of its own. Now and then there are small revelations, but most of the reports focus on other Dutch media, which are not covering the story or do so only minimally. "Why does the NATO press say nothing about Kistemaker?" the newspaper asks after the first revelation. In the article, the silent middle-class press as well as Kistemaker himself, in addition to the FOM institute, come in for pointed criticism. Furthermore, the paper states the suspicion that higher powers in The Hague and Bonn have a hand in this. On October 22, 1960, there is good news from Moscow, at least from *De Waarheid*'s perspective. The state newspaper *Pravda* warns against the consequences of the Dutch-West German ultracentrifuge project.

At the beginning of November, when two other newspapers, the independent *Het Parool* and the Social-Democratic *Het Vrije Volk*, finally write something about Kistemaker, *De Waarheid* features a patronizing lead article the next day that makes mincemeat of colleagues with different political convictions. "They cannot maintain silence . . . but they suppress the facts," it states. *De Waarheid*'s writer states that particularly *Het Parool* has been outrageous. According to this newspaper, Kistemaker is said to have been a Dutch counterspy, sent to Paris by some Leiden professors who did not trust Cellastic. *De Waarheid* does not believe a word of it. "So the young graduate student was allegedly sent to Paris to see how the land lay," is the jeering response.

At the beginning of December, Kistemaker invites the editors in chief of four newspapers to his lab for a closed press conference. Represented are *Het Vrije Volk*, *Het Parool*, *Algemeen Handelsblad* (independent) and *Algemeen Dagblad* (independent), and they hear from the tormented research director that he does indeed do secret research. He would like to know that the gentlemen present will publish nothing that could be against the interests of NATO. He also says that he expects "the Communists" will not be able to maintain the campaign much longer and that it is therefore desirable to dedicate as few words as possible to the whole issue.

Gradually, the issue drives Kistemaker ever more into a corner. Staff members of the laboratory in Watergraafsmeer are beginning to feel uncomfortable about the hubbub in the newspapers and the questions in Parliament; some students, on the other hand, want to push things much further, and among other activities, picket Kistemaker's home. On the fence on the other side of the street from the laboratory, someone paints in huge white letters: "KISTEMAKER = A-BOMB MAKER." To leave no doubt, a swastika has been added at the end.

In December 1960, the goaded Kistemaker writes in a personal letter to Goudsmit that things are getting to him. "The campaign has already lasted two months, and that is too much even for someone with a thick skin." In a late-January scuffle with demonstrators at the front door of his lab, the professor pulls a camera out of the hands of a photographer of *De Waarheid*. The latter lays charges, but a few weeks later the action is dismissed because of lack of evidence. During the same scuffle, a student of astronomy claims to have been hit ("I saw stars"), or almost hit, by a police officer, but this matter is dismissed as well. By this time, *De Waarheid* has already established a benefit fund for the unfortunate student and has interviewed him at length.

In the meantime, *De Waarheid* has worries of its own. The Communist writer covering the story is upset by the suggestion, made in other newspapers, that *De Waarheid* has dug up the entire Cellastic affair from the pages of Sam Goudsmit's book about Alsos. On the contrary, he claims, the facts of Kistemaker's wartime past came from other sources. Alsos was a supplement. This is because Goudsmit is not a good source at all, *De Waarheid* states viciously: "The fact that Goudsmit was and is intimately acquainted with Professor [Wilhelm] Groth, does not exactly increase the trustworthiness of his judgment about this scholar during the Nazi period."

Goudsmit never felt comfortable with the role that his Alsos adventure, fifteen years after it took place, came to play in one of the most contentious affairs in the post-war Netherlands. In 1960, he tries at once to calm people down. After the *De Waarheid* articles, *Het Vrije Volk*, a newspaper allied with the Labor Party, asks the Dutch-American what he thinks of the commotion. "Professor Goudsmit writes this: 'I hope that these unnecessary attacks will not hinder Kistemaker's important scientific work.'"

According to *Het Parool*, Sam even expresses regret over the publication, so soon after the war, of his book's chapter about Cellastic. Whether that was really the case is now impossible to determine, but the Kistemaker affair will continue to follow him in later articles and interviews. "I believe the whole business should now be buried," he answers in 1968 to a Nijmegen history student who wants to write an essay about Cellastic. "Everything I know is described in my book. I believe I really should dissuade you from raking up that old junk again. The most important thing that can be learned from it is that scientifically-trained people, too, can be very weak and stupid. But I believe the world knows that by now!"

Goudsmit repeats his distaste for the Dutch commotion in an interview with the weekly *Haagse Post*. The attacks by *De Waarheid* he considers to be below the belt. It is still his impression that Cellastic amounted to absolutely nothing in the nuclear area. And the Kistemaker issue itself was dealt with by the military intelligence service interrogation and the exoneration afterward: Kistemaker had been foolish but not really a collaborator. To keep raking up that matter makes absolutely no sense, in Goudsmit's view, except perhaps for those who have a present-day political agenda. Besides, what annoys him immeasurably is the persistent suggestion that Kistemaker's obscure work in Paris in the war had anything to do with a potential German nuclear program. "There is not a single reason to believe that Kleiter or anyone else from Cellastic had anything to do with the German nuclear project in the war." He mentions that the well-known British writer and historian David Irving does not devote a single word to Cellastic in his book about Hitler's nuclear research.

But that is all he is willing to say publicly. In the early 1960s, Goudsmit advises Kistemaker in writing to remain calm under the attacks of *De Waarheid*. "Avoid publicity and keep a low profile," he writes. Sam is seriously irritated by the clumsy way in which the gruff Kistemaker defends himself against all the commotion.

During Goudsmit's later visits to the Netherlands, he occasionally meets Kistemaker, but there is certainly no question of warm friendship. The two are too different in character for that. Sam Goudsmit is a man of principle who has, moreover, been marked by life. Jaap Kistemaker, on the other hand, is a pragmatist through and through, who evidently worms his way through everything without much thought or remorse. That does not make for a close friendship.

What Goudsmit really thought about Kistemaker is perhaps

easiest to infer from a personal letter written in June 1967 that showed up in a Leiden archive a few years ago. Goudsmit's letter is addressed to Dolf de Vries, a chemist who has worked for Kistemaker all his life, but who, "especially as a Jew," after all the to-do has begun to doubt his boss's ethical outlook. Goudsmit has known De Vries since the 1950s, when he worked for a while as a post-doctoral fellow in the United States. Years later, De Vries asks Sam for his judgment: did or did not Kistemaker collaborate with the Germans during the war?

Not long before his death, De Vries told a visiting historian how relieved he felt by Goudsmit's answer, however resigned it may sound today: "The best defense, one Kistemaker unfortunately never uses, is that his behavior was extensively examined after the war and that he was exonerated. As usually happens, the least guilty have the greatest feeling of guilt, and that is his problem. He is always on the defensive, and that does not improve the situation."

A year later, Goudsmit writes something similar to the Nijmegen student with the essay request. "It is too bad that Kistemaker feels the need to keep defending himself; besides, he does it in a very foolish manner and does not listen to advice."

In 1971, *Waarheid* journalist Wim Klinkenberg publishes a book about the assumed German A-bomb with the title *The Ultracentrifuge*. In it, he explains once more why he thinks Kistemaker's efforts with respect to Urenco should be understood as nuclear cooperation with those "damned Germans." Klinkenberg continues to insist that Kistemaker has helped the Germans, even though ultimately it is American-NATO bombs that will be stationed in Germany. He is not convinced by Kistemaker's exoneration of the suspicion that he knowingly and willingly worked for the Germans during the war. "That Kistemaker is no good," continues to be the assessment in the offices of *De Waarheid*. The only difference from the pieces written in the early 1960s is the name of the most prominent German authority. Initially that was still Konrad Adenauer. Now it is the very popular and very right-wing Franz Josef Strauss, leader of the conservative Christian Social Union in Bavaria. For the average *Waarheid* reader, no post-war German politician could be scarier than this Strauss.

Wim Klinkenberg, in his own words the last Stalinist in the Netherlands, died in 1995. Even the fall of the Berlin Wall did not change his mind about Communism.

CHAPTER 7

HEISENBERG'S VERSION

Cambridge, 1945–1973

Early in October 1945, Sam Goudsmit returns from Europe to America. The detective work in Germany is not yet done, but the war in Europe has ended, and he is worried because of reports from his old friend George Uhlenbeck that Jaantje is seriously ill. Sam has discussed the situation with Colonel Pash. He has recommended Sam for leave and arranged a flight to America for him. Once at home in Cambridge, it turns out that things are better than anticipated. His wife has indeed had a serious bout of pneumonia, but it has been brought under control. Nevertheless, the reunion with Jaantje and Esther, by now thirteen years old, has an uncomfortable and painful side, now that Sam knows for sure that his parents are really dead, gassed in Auschwitz. He tells Jaantje circumspectly about his terrible visit to the old house on Koninginnegracht in The Hague. About the old school notebooks that were scattered here and there, the broken windows. What exactly he has been doing in Europe goes mostly undiscussed. That is still secret.

In the following weeks, Goudsmit is in Washington, writing one report after another about the work on a potential atom bomb in Nazi Germany, and about the scientists involved. Ten of them are still in custody on a British country estate, Farm Hall. True, the conditions are agreeable. The transcripts of the men's conversations—everyone, from Heisenberg and Gerlach to Von Laue, is listened to with concealed microphones—go straight to the Department of Defense in Washington for analysis. The Farm Hall tapes will remain classified until the 1990s, but Sam reads them all and will use what he reads and hears, together with thousands of documents captured in France and Germany, in *Alsos*. The conversations at Farm Hall strengthen him in the view that he already formulated during the war: that the Germans did not build an atom bomb resulted above all from the lack of interest in science under the totalitarian Nazi regime. The concentration of power and hierarchical thinking smothered all scientific creativity.

In one of the reports—the covering page is marked "TOP SECRET"—Sam assesses the desirability of letting the captured German scientists return to their homeland. His comment about Heisenberg is penetrating: "Heisenberg is a special case. He openly opposed the Nazis and was even called a white Jew. On the other hand, he is very nationalistic, has undemocratic tendencies and is convinced of German superiority. He is, that is certain, humane, and is horrified by all the cruelties of the Nazis. But all in all, it does not seem prudent to allow him to assume a position in which he could work against the interests of the Allies."

Upon his return home, Sam has considerable difficulty in getting back into his pre-war routine. "I have been very busy the last few months," he writes to a friend in London a good six months after his return home, "and have found it hard to get used again to simply teaching and to doing physics. This summer I am temporarily teaching at Harvard, and in September I am joining the Physics Department of Northwestern University in Evanston, Illinois."

After his military adventures in Europe, Goudsmit decides to make a new start in science, but that does not go smoothly. His old position at the University of Michigan has been filled, his work on radar technology at MIT during the early war years was emphatically temporary. After his stint as lecturer in theoretical physics to freshmen at Harvard, he is happy with an appointment in Evanston, a suburb of Chicago. The family moves, and very gingerly family life regains shape in a new house and a new city. Sam has to get used to Esther, who is no longer the young girl who waved goodbye to him in 1943, but a self-conscious teen with adolescent outbreaks. However, with the appointment in Evanston, the continuing money worries are past. As a theoretician and division head, Sam receives a commensurate salary that easily covers family expenses.

The war's conclusion leaves a mark on many of the physicists who were involved in the construction of the atom bomb. Right after the nuclear attacks on Japan in August 1945, many begin to have doubts. When the United States built an atom bomb in order to stay ahead of a potential German atom bomb, which turned out not to exist, it created a new and terrible weapon without needing to. The horrors that appear in the media depicting the destruction of two Japanese cities, where tens of thousands have died, make this realization almost unbearable.

In the autumn of 1945, a stenciled newsletter comes into existence in which concerned atomic scientists exchange news and opinions: the *Bulletin of the Atomic Scientists*, produced in Chicago. Sam is asked to join the editorial board and agrees at once. He, too, is wrestling with the atom bomb. The first pieces he writes for the *Bulletin* are about what he came across in Germany, and why the Germans, although they discovered nuclear fission, did not build an atom bomb. Goudsmit is still constrained by military secrecy; the articles he publishes are first submitted to the Department of Defense.

Sam's name emerges for the first time in the *Annals of the Society of Atomic Science* in January 1946. He figures in a short report on Senate hearings held in Washington at the beginning of December about the development of the atom bomb: "Dr. Goudsmit, who led a mission to Germany to discover what the Germans were doing about atomic energy, revealed that we knew just as little about the German progress as they did about ours. Until it turned out a year ago [end of 1944] that the German atomic project had accomplished little.... According to Dr. Goudsmit, they had no test project, everything was in the laboratory stage. Their scientists had the feeling that it could be fifty or a hundred years before the research yielded something."

At the same hearings in Washington, the physicist Eugene Wigner, later one of the inventors of the hydrogen bomb, says it makes no sense at all to keep knowledge about nuclear energy a secret. That will only obstruct science unnecessarily, Wigner opines, and will ultimately not work anyway. He names three areas that are at present completely secret: the physical constants of bomb materials such as uranium and plutonium, methods of calculation, and technical details of the installations used. The questioning of dozens of prominent nuclear scientists continues into the spring.

In February 1946, Sam uses the pages of the *Bulletin* to amplify his findings in Europe about the German nuclear program. At that time, the emphasis for him still rests on the physics of the nuclear weapon. Crucial, he writes, was that from the outset the German scientists involved in the project did not believe in the achievability of a plutonium bomb. They concentrated completely on uranium, of which the fissionable fraction that is needed for a bomb is difficult to separate from the useless fraction. Without that enrichment with uranium-235, a uranium reactor is thinkable, but a bomb is ruled out.

In mid-March, almost four months after his remarks before the Senate committee, Goudsmit publishes a long article in the *Bulletin*. It is, in fact, a synopsis of his book *Alsos*, which will appear a year and a half later. "How Germany Lost the Race," is the title. The two-page article is an analysis of the riddle that has kept him busy for more than a year: how the discoverers of nuclear fission failed to build an explosive with it, in spite of their head start of at least a year on the Allies. The analysis comes down roughly to what he told the committee in December and is strictly businesslike. The adventure of the Alsos mission is absent from this piece. That will be added later, with the journalist Edwin Seaver as ghostwriter. The book will have a telling subtitle: *The Failure of German Science*. Sam himself sees the political lessons of Alsos as more important than what has actually been discovered about the German nuclear program. They will become ever more determinative for his views about the post-war American science enterprise, which he will observe from up close as editor-in-chief of the journal of the American Physical Society.

"My conclusions are based on having had access to all German material and on having contacted all key men on this project in Germany," he writes. He then explains, point by point, what the Germans and their scientists thought they were doing in the years from 1939 to 1945.

The German line of thought was as follows:

1. An energy-producing uranium engine is more likely to succeed than a bomb. In fact, they had entirely abandoned the hope of making a bomb during this war.
2. An atomic bomb is an [*sic*] uranium engine which gets out of control; therefore the road toward a bomb leads via the construction of the uranium engine.
3. To make a bomb of pure plutonium never entered their minds, or at least was not considered feasible and not taken seriously.

The idea of using a pile to produce plutonium and to make a bomb out of that material came to them only slowly, after the detailed radio descriptions of our bomb in August 1945.

4. An uranium engine is just as important as a bomb because it will make Germany economically self-supporting by the enormous power it may produce.

The remarkable thing, Goudsmit writes, is that throughout the war the Germans thought they were far ahead of all competitors. Only after the American atom bombs fell on Japan in August 1945 did they realize how terribly far they were behind.

Goudsmit thinks the physicists themselves cannot really be blamed: the suffocating Nazi regime handicapped science. At the Armaments Ministry, for example, the physicist Erich Schumann, grandson of the composer and a prominent Nazi, simply did not believe the promise of nuclear fission. Moreover, there were several different departments that worked separately on the uranium problem, without consulting each other. That lack of coordination was corrected only in 1943, when the Germans began to lose the war, and by then it was much too late. Science, Goudsmit also notes, did not exactly hold a favored position under the Nazis. Initially, physicists, whether talented or not, were conscripted for military service just as others were. A number of them even died in battle.

By the end of the war, Goudsmit notes, the Germans knew from their experiments that augmentation of the liberated neutrons in a nuclear reaction was possible, but they certainly did not have a self-maintaining source of neutrons (necessary for a reactor or a bomb). Nevertheless, the German nuclear scientists thought that this insight would play a role in the negotiation of conditions for a peace after the war. And what was more: not only would the German route to atomic energy prove interesting for the world, but it would also put German science on the map after the war. "These thoughts were, indeed, the driving force behind the German scientific efforts."

Goudsmit estimates in his article that altogether the Germans put around ten million dollars into nuclear research. He also thinks that in all those years at most a hundred scientists worked on it. Both are a small fraction of what the Americans and Allies devoted to their Manhattan Project.

In the last paragraphs of his piece, Goudsmit sums up the lessons to be learned from the German nuclear failure. The passage can be read as a creed that will govern a large part of his post-war life as a prominent champion of scientific freedom. Keeping scientific knowledge secret is dangerous, he believes, in any case more dangerous than openness. "I feel certain that, if all countries grant complete scientific freedom to their research workers—the scientists' free choice of research problems and freedom of publication—no dangerous activity will, or can, be kept secret as long as scientists are involved."

Noteworthy, too, is that nowhere in his article does Goudsmit discuss potential moral questions, not on the German side and not on the American side, either. For him, the story of the German atom bomb is primarily one of an accumulation of misunderstandings and mistaken assessments, not of internal resistance to working on a superweapon. In any case, the Allied side also had its share of misunderstandings. The opening sentences of his book *Alsos* are striking: "Looking back to the end of 1942 and the early months of 1943, from the vantage point of the information we now have, one must admit that there was an element of high comedy in the state of mind that prevailed at that time among American and German scientists. The Americans, having succeeded in producing the first chain reaction in a uranium pile, and seeing the atom bomb as a definite possibility, were certain that the Germans must know as much and more. . . . They had begun their uranium research two years before ours. And especially, and above all, everyone knew that German science was superior to ours. The Germans were just as positive that their science was superior to ours. To be sure, they were still groping in the dark, but, they figured, if they were in the dark, where must the Americans be?"

Sam Goudsmit himself already realized during the war that the atom bomb would be built and used, even after it turned out that the Germans did not have one. At the end of 1944, after the liberation of Strasbourg, Sam concluded that the Germans had not developed a nuclear weapon. But when he spoke about this to the Mysterious Major with a sense of relief, and said that now the American bomb would not have to be used, the more realistic major set him straight: "Of course you understand, Sam, if we have such a weapon, we are going to use it."

At the beginning of 1946, Werner Heisenberg is released from Farm Hall and taken to Göttingen, which is still under British authority. There he first hears about Goudsmit's plan to write a book about the failure of the German atomic scientists. During the past year, Goudsmit has repeatedly given his view of the matter in articles in the *Bulletin of Atomic Scientists*. Aside from the Nazis' suffocation of science, Goudsmit thinks, the German scientists had only half understood the physics of a nuclear bomb.

Heisenberg has read those articles and his scientific honor has been cut to the quick by the suggestion that he, Nobelist and founder of modern quantum physics, may have made an error concerning nuclear fission. That is not how it was, he thinks. Even before the appearance of Sam's book *Alsos*, Heisenberg publishes a long article in *Naturwissenschaften* in which he gives his version of the matter. The article also appears in translation on August 16, 1947, in the British scientific journal *Nature*.

Heisenberg's article opens with a remarkable paragraph about the discovery of nuclear fission. It is true, he writes, that the fission of uranium nuclei into two equal portions was discovered in December 1938 by two German researchers in Berlin: Otto Hahn and Fritz Strassmann. But it was a Frenchman, Frédéric Joliot-Curie, who saw in the spring of 1939 that in fission, neutrons are liberated that are potentially able to split other uranium nuclei. "Thereafter the possibility of nuclear chain reactions was eagerly debated among physicists, particularly in the United States." The German researchers Lise Meitner and Otto Frisch were the first to calculate, also early in 1939, that potentially a lot of energy would be liberated through fission. In his strictly historical account, Heisenberg does not mention that Meitner had already fled to Denmark.

In spite of the German roots of nuclear fission, Germany historically had very little interest in the atom. In the 1930s, a series of laboratories in the United States, France, and England were equipped for nuclear research, often with their own cyclotrons. During that time, nothing much happened in Germany in that field. The government barely concerned itself with physics research. Nuclear laboratories were built only in Heidelberg and Berlin-Dahlem, financed by the Kaiser Wilhelm Institute. In Heidelberg, the construction of a cyclotron, an accelerator with which atoms can be selected, was commenced in 1938. But it would be tested for the first time only in 1944.

According to Heisenberg, reports reached Germany at the beginning of the war that the American military authorities had an uncommon interest in nuclear research. "In view of the possibility that England and the United States might undertake the development of atomic weapons," Heisenberg writes, "the *Heereswaffenamt* created a special research group, under Schumann, whose task it was to examine the possibilities of the technical exploitation of atomic energy." Men such as Diebner, Bothe, Clusius, Döpel, Hahn, Harteck, Joos, and

Von Weizsäcker were quickly recruited into this *Uranverein*, uranium association.

The Kaiser Wilhelm Institute for Physics in Berlin-Dahlem became headquarters, under the auspices of the Armaments Ministry, the *Heereswaffenamt*. To this point, the director was Nobelist Peter Debye. But as a Dutch national, Debye was not allowed to work under the *Heereswaffenamt*, so he had to step down. He declined naturalization. In 1939, he left for America, officially for a lecture tour. He never returned to Germany, although his family remained in that country. (In the United States, Goudsmit among others regarded the arrival of the German Dutchman with eagle eyes. What was this top scholar coming for?)

Starting in the autumn of 1939, Schumann's *Uranverein* went looking for possibilities of using nuclear energy, as Heisenberg neutrally puts it, "either [for] the controlled production of energy, . . . or directly as an explosive for bombs." A year later, the first pile reactor was built for research: a small tower of slices of uranium oxide and light paraffin. As expected, no chain reactions occurred in it. "Nevertheless [the experiment] yielded valuable data for further using alternate layers of uranium oxide and heavy water."

Around that time, Carl Friedrich von Weizsäcker, who in 1938 was one of the first to understand the possibility of nuclear explosives, made an important theoretical discovery. With the nuclear reactions in a uranium pile, atoms of a new, heavy, uranium-like element will automatically come into being that must have approximately the same fission capacity as fissionable uranium-235. The advantage of this new element, uranium-239, is that it will probably be chemically distinguishable from uranium. Therefore, it is easy to procure in pure form by a chemical route. In view of the difficulty that the Germans had with finding a method to separate the rare uranium-235 from natural uranium, this was an attractive idea.

But they could not really figure it out, Heisenberg writes, looking back. "Since no cyclotron was available in Germany, these elements could not be prepared in sufficient quantity for the examination of either their nuclear properties or chemical characteristics."

The lack of opportunities for studying the new element had the consequence that German nuclear-fission research automatically directed itself to uranium reactors as a source of heat. For this, pure fissionable uranium was unnecessary; ordinary uranium also sufficed. At the same time, Heisenberg describes in *Nature* how German science essentially just kept on working during the war. Scientific questions were studied and sorted out with scientific exactitude, there were laboratories and equipment, a lot was published, also in the regular scientific journals. Heisenberg's account of nuclear research during the war sounds remarkably systematic and well-considered, as though German scientists did not want to be distracted and diverted from their course, no matter what was happening.

Heisenberg writes that he realized in 1940 that an energy-producing reactor could automatically yield material for a nuclear explosion, in any case in theory. Only the means to research this were absent. The Americans, on the other hand, opted for another route; with uranium reactors, they made enough plutonium for their own atom bombs.

The central point of Heisenberg's account, then, is not physics itself, but the conditions under which the German nuclear scientists had to work. Under the circumstances of the war, German atomic research did not have a real chance, he believes. "[The uranium project] could not have succeeded on technical grounds alone: for even in America, with its much greater resources in scientific men, technicians, and industrial potential, and with an economy undisturbed by enemy action, the bomb was not ready until after the conclusion of the war with Germany."

There was, Heisenberg writes, no personnel, no material, no

industry that could complete the uranium project. Moreover, Germany had neither time nor patience: "... The men responsible for German war policy ... expected an early [end] of the War, even in 1942, and any major project which did not promise quick returns was specifically forbidden. To obtain the necessary support, the experts would have been obliged to promise early results, knowing that these promises could not be kept. Faced with this situation, the experts did not attempt to advocate [to] the supreme command a great industrial effort for the production of atomic bombs."

In fact, Heisenberg concludes, the Germans did not have to choose whether or not to build an atom bomb. They lacked the possibilities to do so. "The circumstances shaping policy in the critical year of 1942 guided their work automatically towards the problem of the utilization of nuclear energy in prime movers." What was more, that seemed like an important task for after the war as well. "The mere possibility of solving the problem had been [accomplished] by the discovery of the German scientific workers Hahn and Strassmann, and so we could feel satisfied with the hope that the important technical developments, with a peacetime application, which must eventually grow out of their discovery, would likewise find their beginning in Germany."

When Goudsmit gets to read Heisenberg's version of the history of the atom bomb in *Nature*, he is furious. Above all, he is enraged by the suggestion that the German scientists consciously did not work on a nuclear weapon while the Americans did. That is not the way it happened, he will expound in a long series of articles and speeches. The Germans did not obstruct the construction of the bomb; they did not understand the necessary physics of the atom bomb well enough, and they were unable to explain to their short-sighted superiors how you go about building an atom bomb.

A month after his article appears in *Nature*, Heisenberg writes a personal letter to Goudsmit ("*Lieber* Goudsmit"). In it he explains that he could no longer stand by while Sam systematically described German nuclear research as incompetent, and so he felt compelled to write the piece that appeared in translated form in *Nature*. According to Heisenberg, there never was a race toward the atom bomb between the Allies and the Germans. If a mistake was made, it was made by the Germans. They assumed that the Allies would never begin nuclear research and the construction of an atom bomb during the war. But there was nothing wrong with German science.

Goudsmit waits for a while before responding, rereads archived documents, searches his own mind. Has he gotten it wrong after all? His answer appears in the November 1, 1947, *Bulletin*. His tone is caustic; he does not yield an inch: "[Heisenberg's] article is a translation of a report intended for *Naturwissenschaften*, a circumstance that may well explain some of the surprising features of Heisenberg's discussion. It has all the earmarks of being meant for German home consumption and home appeasement."

Heisenberg's version is a false success story, Goudsmit emphasizes. "The German public is led to believe that their scientists were on their way toward reaching the goal [of an atom bomb] just as the Allies were. It is true that some of the German experiments were of high quality. . . . But they were on the wrong path as far as the bomb was concerned. It is not right to blame this on a ruling by Armaments Minister Speer or on the war conditions in Germany."

In subsequent years, the failure of the German nuclear project continues to occupy the two physicists almost obsessively. They write each other a series of frosty letters with arguments and reproaches that end with Goudsmit's proposal to stop writing about it altogether. Not only has the discussion bogged down in an endless succession of moves and counter-moves, but Sam thinks the public is rather fed up with the Second World War. The world is now caught up in the Cold War. There are bigger worries than who said what to whom in 1944.

In 1947–48 it has not yet gotten that far, however. On August 20, 1948, Goudsmit writes a letter to Heisenberg in which he uses sharp words to describe how painful he finds the situation. "Writing this letter is almost harder than solving a big problem in physics,"* Sam writes dejectedly. He sticks to his view that Heisenberg would do better to acknowledge that German physics simply failed. They just kept messing around with uranium.

On September 23, Heisenberg answers from Göttingen. The choice of uranium reactors seemed logical, for energy generation, that is. A bomb was too complicated, possibly also for the Americans, he opines. "I have spoken about the matter again with Bohr. We never seriously sought about a race with the Allies. A situation like the one in which we found ourselves leads to a more passive and modest approach."*

On December 1, 1948, Sam has worked up the spirit to respond to Heisenberg. "It is clear that we hold different opinions. I was surprised and shocked by your article in *Nature*. If the Germans had known about plutonium in 1940, you would surely have worked in that direction."*

And the matter gets worse. Four weeks later, on December 28, 1948, the *New York Times* published the very first major interview with Werner Heisenberg about his role in the war. The piece is written by science editor Waldemar B. Kaempffert. In the interview, Heisenberg sticks to his view that the Germans were conscious of the possibility of a nuclear weapon, but never wanted to take it up fully. Besides, the dismal circumstances of the war made that all but impossible. "Fortunately," said Dr. Heisenberg, "they never had to make a moral decision, and this for the reason that they and the Army agreed on the utter impossibility of producing a bomb during the war."

Sam spends the week after that interview in his office at Brookhaven, surrounded by his voluminous archive. With the knowledge of the War Department in Washington, he has saved files full of original or copied German documents from the war, from the Strasbourg documents that in his view showed there was no German bomb project, to the transcripts of the German discussions in Farm Hall. He wants to go through the documents once more; he has decided that the rejoinder to Heisenberg's version has to be tougher, more factual.

On January 14, 1949, Goudsmit replies to Heisenberg. "I was still engaged in gathering material for a reply to your letter of December 1, when the article in the *New York Times* appeared. In that article, you raise erroneous impressions. Your need to save the prestige of German science is totally out of place. The documents that I have certainly do not put the German work in the war in a favorable light. These documents are at this moment still officially secret, but that may be changed. For these reasons, it is probably better simply to cease these discussions."

A subplot of the discussion is Goudsmit's deep disenchantment with his old colleague Heisenberg. He had thought better of him than the twisting and turning with which he emerged from the war. He finds him pedantic and untruthful, and that is unworthy of the real scientific hero that the superior physicist Heisenberg undeniably is.

Much later, Goudsmit speaks about Heisenberg to the German historian of science Armin Hermann. There was, Sam says, a certain unconscious disappointment in his anger: "I realized more and more that this great man was really was no wiser than most of his colleagues and had shown no humane leadership."

It is difficult not to look up to Heisenberg as a theoretical physicist, the man who as a boy almost singlehandedly gave an entirely new turn to quantum physics. Sam, on the other hand, sees himself as a bit of a lucky misfit in theoretical physics. His discovery of the electron spin in 1925 he regards as a fluke. Probably he is selling himself a bit short as an academic, but compared to Heisenberg, he is indeed a minor scientist. In the 1920s, when Goudsmit, a hesitant student of Ehrenfest in Leiden, gets bogged down in a theoretical analysis of the helium atom, it is the equally young Heisenberg who casually solves the problem. "That was beyond me. I would never have been able to guess at anything like it," Goudsmit confesses later, with a hint of envy of so much effortless talent.

On the other hand, the friendship was never wholly free of mutual amazement and incomprehension. For example, at the beginning of April 1936, Heisenberg sends a note to his colleague Sam Goudsmit in Ann Arbor. He cannot come to the university's summer school after all, he reports, because he has to do a period of military training. Heisenberg is also a corporal in a Bavarian mountain unit, posted near the Austrian border. In letters to his mother that summer, Heisenberg repeatedly writes how much he likes being there, even if he does not have much time for physics. Goudsmit is surprised that a genius like Heisenberg can consider mountain hikes and other boyish activities to be more important than physics.

Ultimately, time performs miracles. In September 1950, Sam and Werner Heisenberg meet face to face for the first time since the initial interrogations in Heidelberg in 1945. Heisenberg, who went back to work in Göttingen as a professor in a laboriously recovering Germany at the beginning of 1946, has traveled to the United States for a professional visit. Before going, he has cautiously sounded out Goudsmit whether a meeting is feasible. To that end, he has written a soothing letter to Sam on June 22, 1950. "You can rest assured that I for my part will not resume the disagreeable discussions of years past. I look forward eagerly to exchanging thoughts about interesting developments in physics."*

The meeting takes place in Brookhaven, where Heisenberg gives a lecture to the members of the Physics Department. The reunion of the two men is cordial, but not warm. "The friendship of the olden days was gone," Sam's daughter, Esther, says later. Heisenberg stays in a hotel in the neighborhood, not at home with the Goudsmits as in the 1930s.

All the same, changes do occur in the course of the years. Sam judged Heisenberg harshly in his report about an eventual release from Farm Hall ("it does not seem prudent to allow him to assume a position in which he could work against the interests of the Allies"), but he is very mild five years later. In a 1952 letter to the Atomic Energy Commission (AEC), the subject is a potential appointment in the United States for Heisenberg, who has already made a successful professional visit to America and also visited Sam in Brookhaven. Goudsmit sees no objection to appointing the German theoretician in the United States, he writes, especially not if that is a way of ensuring that he does not fall into the hands of the Russians. Sam does wonder whether after the war Heisenberg is sufficiently in touch with current theoretical physics.

Still, a position with influence on policy strikes Goudsmit as a bad idea. "He does definitely have less admirable character traits and went too far in his loyalty to the former Nazi regime," he maintains.

However, Sam has already indicated that it is time to bury the tomahawk. "Before we lose our self-control, a final letter about this controversial subject. I will summarize my position one more time. The lack of progress of the German scientists during the war was entirely due to the suffocating atmosphere of a dictatorial regime. That matters more than the deficient material circumstances in wartime. Above all, I had hoped that you in particular would have told the world about the importance of academic freedom. Nobody is better qualified to do so than you are.'" And as if to break the ice even further, Goudsmit abruptly starts to write about recent discoveries in the field of ion physics.

Werner Heisenberg publishes his memoirs in 1971. In them, he accuses Goudsmit of not giving the full picture, at least not in his 1947 book *Also*: "There are countless things that Goudsmit failed to see, whether deliberately or not." Goudsmit ignores this late swipe; as far as he is concerned, the time for heated debate is past.

In April 1973, Heisenberg is in America for public lectures as part of the Copernicus Year. He meets Sam in a Washington hotel on April 23. The two men spend the evening over cocktails and dinner, to their mutual enjoyment. It is the last time they see each other. (Heisenberg dies in Munich in February 1976, almost three years before Goudsmit.) For thirty years, Goudsmit was unable to bring himself to ask about that 1943 letter sent to his friend and colleague Dirk Coster, in which Heisenberg coolly stated that he hopes nothing bad will happen to Sam's parents. At the time, Isaac and Marianne had already died in Auschwitz. For his part, Heisenberg never touched on the topic again, but he must have known what happened.

In 1973, too, German television asks Sam to take part in a live debate about nuclear weapons with Carl Friedrich von Weizsäcker and the father of the American hydrogen bomb, Edward Teller. Sam has his doubts. He writes to Heisenberg: "I believe our differences of opinion are much less important than some outsiders think. I hope that we will speak with each other again and determine that we are actually of one mind, within the confines of the spoken word. If you are planning to come to the United States, do let me know so that we can meet each other.'"

In his reply, Heisenberg agrees that for outsiders the discussion may seem more hostile than it is. Quite frankly, he finds it rather excessive that the television people want Sam to fly to Europe for a public discussion. "I would not do it and do not find the discussion that important. On the other hand, it will harm nobody to hold it." The debate does not take place.

While Heisenberg and Goudsmit are letting the matter rest at last, the rest of the world is far from finished with it. In 1956, a book finally appears about the history of the atom bomb, *Brighter than a Thousand Suns*, by the Swiss journalist Robert Jungk. It is based in part on conversations with Heisenberg. Jungk follows Heisenberg's version, in which the German nuclear scientists forgo an atom bomb on practical grounds and concentrate on a uranium reactor. Central in the account is Heisenberg's famous 1942 visit to Niels Bohr in Copenhagen. According to Heisenberg's account, Heisenberg tried to reassure his friend and mentor Bohr: "the Germans are not building a bomb." Bohr later remembered the conversations very differently: Heisenberg was bragging about the advanced German nuclear research. The misunderstanding probably arose because Heisenberg did not explicitly use the word "bomb." In German eyes, that would have been high treason, something from which, as a nationalist, Heisenberg recoiled.

We cannot know who is right about the ability or inability of German science to build an atom bomb. It might have gone differently than either Goudsmit or even Heisenberg knew in the 1950s. Many years later (in 2005), the German historian Rainer Karlsch published *Hitlers*

Bombe. That book again raises the intriguing question of whether Sam Goudsmit and his Alsos mission really got a complete picture of the German efforts to develop an atom bomb. In archives and statements by witnesses, Karlsch found indications that, parallel to the scientific work of Heisinger, Diebner, and Gerlach, there had been a second nuclear program in Nazi Germany. That, Karlsch claimed, had possibly even led to tests with explosives that carried a nuclear charge.

On March 4, 1945, an extremely violent explosion occurred in Ohrdruf, Thuringia, after which soldiers from the surrounding area were commandeered to take away hundreds of corpses and destroy them. Many had ghastly burn wounds, a helper remembered. Possibly they were prisoners of war who had been exposed to an experiment. And there was more. Years after the alleged event, a sometime prominent Italian journalist claimed that after a visit to Hitler in Berlin, he had been taken to the North Sea island of Rügen, where he was allowed to witness a very violent test explosion. A nuclear explosion, he suspected later.

It did become apparent that soil samples taken from both locations half a century later provided no real evidence of a nuclear explosion. Furthermore, huge question marks surround the reliability of the witnesses' declarations that were used. Still, it is striking that incidents and locations come to light in Karlsch's book that were missed by Goudsmit's men. Ohrdruf and Rügen do not show up in the Alsos archives.

The question remains of how important that really is. The fact is that during the war the German were unable to use nuclear weapons. What *is* interesting is the question, asked with the benefit of hindsight, of whether the choice of Goudsmit for the Alsos mission was in all respects well-advised. At the time of the request to lead the mission in 1943, Goudsmit himself had seriously questioned why he of all people had been selected for the job. In the end, he decided it was because of his connections in Europe and especially in Germany, as well as his

Sam Goudsmit *(front row, left)* with fellow graduate students and faculty members in Leiden, ca. 1925. Albert Einstein is standing in the open door; Paul Ehrenfest is second to Einstein's left.

Photo © AIP Emilio Segrè Visual Archives, Goudsmit Collection.

Paul Ehrenfest and associates, ca. 1927. *From left to right*: Gerhard Dieke, Sam Goudsmit, Jan Tinbergen, Paul Ehrenfest, Ralph Kronig, and visitor Enrico Fermi. *Photo © AIP Emilio Segrè Visual Archives, Fermi Film Collection, Goudsmit Collection.*

The Zeeman Laboratory on Plantage Muidergracht in
Amsterdam, where Goudsmit worked as a research assistant.
Photo from the author's collection.

Jaantje and Sam Goudsmit, Ann Arbor, 1929.
Photo © AIP Emilio Segrè Visual Archives, Goudsmit Collection.

The RMS *Baltic* arrives in New York harbor, 1920s.
Photo by the Detroit Publishing Company.

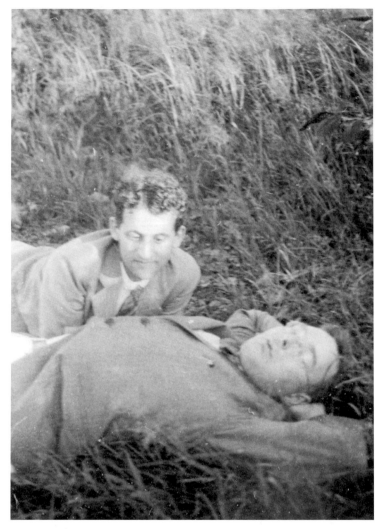

Hendrik Kramers and Sam Goudsmit relax in Ann Arbor during the 1930 summer school. *Photo by Sam Goudsmit © AIP Emilio Segrè Visual Archives, Goudsmit Collection.*

Werner Heisenberg, 1934.
Photo © AIP Emilio Segrè Visual Archives, Bainbridge Collection.

Summer school, Ann Arbor, early 1930s. *Front row, from left to right:* Else Uhlenbeck, Paul Ehrenfest, Jaantje Goudsmit, Enrico Fermi; *back row, from left to right:* Hendrik Casimir, Bart Bok, George Uhlenbeck, Sam Goudsmit, Gerhard Dieke. *Photo © AIP Emilio Segrè Visual Archives, Goudsmit Collection.*

Werner Heisenberg as guest of honor at the 1939 summer school in Ann Arbor. *From left to right:* Sam Goudsmit, Dean Clarence Yokum, Werner Heisenberg, Enrico Fermi, Ralph Kronig. *Photo © AIP Emilio Segrè Visual Archives, Crane-Randall Collection, Goudsmit Collection.*

Sam Goudsmit in a jeep with Lieutenant Toepel, Stadtilm,
Germany, April 1945. *Photo by Malcolm Thurgood © AIP
Emilio Segrè Visual Archives.*

March 1945. The Allies have marched to the Rhine at Mannheim.
Photo by H. R. Weber and Sperry, United States National Archives.

Goudsmit at Alsos headquarters in Heidelberg, April 1945. *Photo © AIP Emilio Segrè Visual Archives, Gift of Michaele and Terry Thurgood, Thurgood Collection.*

The Alsos team dismantles a "uranium machine" in a cave near Haigerloch, Germany, April 1945. *Photo from Brookhaven National Laboratory © AIP Emilio Segrè Visual Archives, Goudsmit Collection.*

The Alsos team removes a hidden supply of uranium in a field near Haigerloch. Sam Goudsmit is seated; to his left is Michael Perrin. *Photo by Samuel Goudsmit © AIP Emilio Segrè Visual Archives, Goudsmit Collection.*

Sam Goudsmit poses with an M20 machine gun in
Stadtilm, Germany, April 1945. *Photo © AIP Emilio Segrè*
Visual Archives, Goudsmit Collection.

Destruction in The Hague after the bombing of the
Bezuidenhout arca, spring 1945.
Photo © Hollandse Hoogte.

Jacob Kistemaker in the IKO laboratory, Amsterdam, 1961.

Photo © Hollandse Hoogte.

Sam's second wife, Irene Bejach Goudsmit, at a reception in
New York, 1980. *Photo from AIP Emilio Segrè Visual
Archives,* Physics Today *Collection.*

Sam Goudsmit, editor of *Physical Review Letters*, ca. 1965.
Photo by Heka Davis from AIP Emilio Segrè Visual Archives,
Physics Today *Collection.*

multilingualism. Besides, he had long had a certain interest in detective work.

All that certainly did Alsos a lot of good. Like no other, Goudsmit was able to discern the thought processes of his former friends and colleagues and to sift through their archives without problems. His detective sense definitely proved helpful during inspections. And his talent for finding patterns in the tangle of data and indications in atomic spectra was important in the intelligence work, too.

Anyone who wanted to know what the Germans were up to in the nuclear field had to get on the trail of the nuclear scientists. From the start that was the logical idea for Alsos, based in part on the experience of the Manhattan Project in America. Goudsmit could fully agree with that. As a researcher, he knew that German science had long been very hierarchically organized. He felt it was impossible that lesser physicists could have had the starring role in research that seemed so new and important, or that military men, for example, would have taken the lead in research.

Alsos's strictly scientific interest was the reason that the routes taken in 1944 and 1945 by Pash's intelligence troops led mainly through European universities: Paris, Strasbourg, Heidelberg, Hamburg, Berlin. That the ultimate dénouement of the hunt for Hitler's uranium program led to unknown villages such as Haigerloch and Hechingen was because that was where the scientists whom Goudsmit knew well were staying, having fled from Allied bombing of the cities. Surely, they had to be the driving forces in the German atomic project?

For Goudsmit, the lesson of Alsos is above all that in the years 1939–1945 the Germans squandered their nuclear lead because the totalitarian Nazi regime suffocated science and scholarship. He will maintain this until his death in 1978. In the post-war decades, he develops into an ever more critical opponent of excessive government control of science. During the Cold War, while American physics departments at many universities are increasingly feeling the hot breath of military men and politicians on their necks, he moves to the American Physical Society and becomes editor of the leading journal *Physical Review*. In 1956, he starts the pioneering weekly *Physical Review Letters*, which will become one of the top publications in science, comparable to *Nature* for many other fields. As editor, he can occupy himself with pure science at last.

And yet background and education will tell. In many a commentary, Sam will defend the interests of scientific and academic freedom. The old war is pretty well past. A new war is in progress, a cold war this time, a conflict in which once more he urgently wants to play a part.

CHAPTER 8

SAM'S COLD WAR

Brookhaven, 1947–1978

In the summer of 1960, Sam and Jaantje divorce after thirty-three years of marriage. Their wedding in 1927 shaped the beginning of their big American adventure, and their marriage brought them a lot. But now it is finished. Sam is somber to the point of depression, struggling with the murder of his parents by the Nazis and the threat of nuclear war. The uncomfortable thought keeps gnawing at him that he did not try hard enough before the war to bring his father and mother to America, away from a dangerously smoldering Europe. They did not want to leave, attached as they were to The Hague and the Netherlands, although they did visit Ann Arbor for a few weeks in the 1930s after Esther's birth. But above all, Sam blames himself for what happened. He has the feeling that he was too lax in applying for entry visas in 1939. Ultimately, he did apply. But it was too late.

Sam does not even know for sure whether his parents ever received the visas after they were issued. The American consulate was undamaged in the bombing of Rotterdam in May 1940, but in the chaos of war, the papers probably never arrived in The Hague. In their farewell letter from Westerbork in late 1942, they did not say a word about it. After the war, when he is asked to enter personal details about his parents in government documents, he enters their date of death as February 11, 1943, when they were gassed at Auschwitz, and the cause of death simply as "Nazis."

Jaantje is not doing well, either. What exactly is the matter is not discussed. Daughter Esther remembers that her mother was in a clinic several times during the late 1950s, but she does not know what for. At the time she was studying in Ann Arbor. When she comes home for visits, Jaantje is flat and emotionless. Her friend Else Uhlenbeck writes to Esther that her mother is receiving calming injections in the clinic at that time. Only after several years, and after the divorce, does she seem to recover some of her old enjoyment of life. In the same letter, Else writes about her occasionally uncomfortable feelings around Jaantje over the years: "She was always a bit distant, almost never talked about herself. I was from a talkative family, so she must have made a rather negative impression of me." Only later does she understand that Jaantje had a difficult childhood, with a mother who died young. Not yet ten years old, Jaantje had to take charge of the household and raise her younger brother. That, she writes to Esther, is one of the reasons she has always insisted on the best schools and colleges for her own daughter. But just as with Sam, the war played a role. Her brother was deported during the war and murdered by the Nazis. "That tore her up inside," Else writes.

In the 1950s, Sam immerses himself ever deeper in his work. Aside from his job as head of the Physics Department in Brookhaven, he has taken a position in the American Physical Society, the country's professional association of physicists, with thousands of members and a head office near Washington, DC. Initially, he becomes editor of the leading scientific journal *Physical Review*. Quite soon he assumes full responsibility as publisher and editor-in-chief. He will continue in these roles into the 1970s. During all those years, the editorial office is located in Brookhaven National Laboratory in Upton, Long Island.

In the 1950s, too, Sam meets an inspiring new love, the much younger Irene Bejach. She is a German-Jewish refugee who ended up in the United States via England as a child. Sam is pushing sixty by now; Irene is in her very early thirties. They are a generation apart, but the war has made them soulmates.

Irene Bejach is the daughter of Jews from Berlin. In September 1939, she arrived in Cambridge, England as an eleven-year old girl with her sister, Helga, two years younger. The two came with the so-called *Kindertransport* that the British government organized to evacuate ten thousand children from Germany before it was too late. The two girls ended up in the family of legal scholar Frederick Levi Attenborough and his wife, Mary. They already had three children, of whom two would become world famous, David as a biologist and nature film-maker, Richard as an actor and director. The latter told the *Daily Mail* in 2009: "I will never forget when Helga and Irene first arrived at our home. They were two pale waifs with their pathetic little cases . . . They looked sad and ill. They were also nervous wrecks."

After the war, Irene was taken into the family of an American uncle, a physician. She trained as a nurse in New York and was working as a medical assistant in New Jersey and on Long Island when Sam met her. They married in 1960, soon after his divorce from Jaantje. They moved into a low-slung wood house on Bayport Avenue in Bayport, New York. Just like his previous house, this place was very close to the huge nuclear lab on Long Island.

A bitter Jaantje leaves for Cambridge, Massachusetts, where she lived with her daughter, Esther, during the war and before Brookhaven. Esther is already studying biology at the University of Michigan, where she writes for the university paper. She will later become a professor specializing in the hormonal regulation of mollusks.

Sam Goudsmit has been working at the Brookhaven Laboratory since 1947. It is a new, safe harbor after he has roamed a bit through the university world, restless and unsure about what he should do with his life as a theoretical physicist. After a short stay at Harvard, Northwestern University is his source of income for two years. But then he transfers to the new lab that will be built on Long Island in the aftermath of the Manhattan Project.

After the war, the United States has a network of national nuclear laboratories that were involved in the Manhattan Project, among other places in Los Alamos, Hanford, and Chicago, but none of them is on the northeast coast of the country. Radar physicist and later Nobelist Isidor Rabi of Columbia University, with his former student Norman Ramsey, begins a lobby for a northeastern nuclear laboratory. In spite of strong opposition from MIT and Harvard, Rabi has rapid success, especially because he has Leslie Groves, the sometime military leader of the Manhattan Project, in his corner. Groves supports the idea that there must be one common American nuclear lab.

The lobby succeeds. At the beginning of 1946, representatives of nine universities come together to establish Brookhaven National Laboratory in the New York area. The owner is the Atomic Energy Commission, the civilian successor of the Manhattan Project. The lab will concern itself with nuclear research and particle physics, and later also material chemistry, neuroscience, energy, the environment, and even issues of peace. It is operated by Associated Universities, Inc., but after a series of incidents with radioactive leaks and fires it will be taken over by the Department of Energy in 1998.

Brookhaven National Laboratory is set up in 1947 on the site of the former US Army Camp Upton, which served as a training center for recruits in both world wars. On March 21, 1947, the complex with barracks and drill grounds is transferred from the Department of Defense to the Atomic Energy Commission. The lab constructs the first civilian nuclear reactor in the world. In the 1950s, it develops one of the first large particle accelerators, futuristically named the "Cosmotron." These are projects in which Goudsmit is only peripherally and mainly organizationally involved, certainly after he assumes the leadership of *Physical Review*. Nowadays, Brookhaven is also home to a large synchrotron accelerator, which is used for X-ray examination at the molecular level. In the course of the years, Brookhaven scientists win five Nobel Prizes for physics and two more for chemistry.

According to Rabi, the coming together of universities in a regional lab is an important prerequisite for peace and stability among countries. In the 1950s, that idea is imitated in a reviving Europe. In 1952, eleven countries, among them the Netherlands, establish their own large particle lab: CERN, the European Organization for Nuclear Research in Geneva.

Goudsmit's 1947 move from Northwestern University to Brookhaven did not happen without a struggle. In a protracted exchange of letters, the university tried to hold him to the terms of a multi-year contract. But Sam did not budge. The work in the new nuclear lab on Long Island was not only more challenging and of greater national importance, but he also had a personal motive. Especially Jaantje was not exactly happy with the move from a heartwarming Ann Arbor, and after that Cambridge, to Evanston, provincial and chilly, a suburb of Chicago on Lake Michigan. After Sam got his appointment in 1946, his family moved briefly to Evanston. But very soon Jaantje and Esther returned to Cambridge, Massachusetts, a much friendlier and culturally more developed place, where Sam spent the early war years in the radar lab of MIT. Esther Goudsmit completed her secondary schooling there.

Brookhaven, on Long Island close to New York, initially feels to Jaantje like a real step forward, a fresh start after the depressing war. Sam, however, still feels it in his bones. In the summer of 1948, the

Goudsmits move to Upton on Long Island. At first, they occupy a wooden shack on the property of the lab that is under construction, pioneers amidst empty grounds, farmer's fields, and extensive forests. At night it is pitch dark, but on the western horizon the sky glows above New York, the city that proverbially never sleeps. On weekends the family makes day trips to the sea, by car half an hour to the south. And it gets better. After a year, Sam and Jaantje buy a house for the first time in their lives, in the sleepy coastal village of Sayville. It, too, is on Long Island, and less than twenty minutes by car from the rapidly rising lab.

The detached wooden two-story corner house is at 78 Benson Avenue, a calm two-lane street flanked by detached houses with driveways and expansive front lawns. The street ends at the beach half a mile farther on. There is a bay with a small yacht harbor.

Small-town America is idyllic. The air is clean, the tall clouds rising above the sea are often spectacular, and Sam and Jaantje become part of their neighborhood and celebrate birthdays, Thanksgiving, and Christmas with their new neighbors. Jaantje works in the local public library. During the day, Sam is at Brookhaven or away traveling, but now and then he lectures in the area about his hunt for Hitler's atom bomb. They will live here until their divorce in 1960. In the meantime, Esther is studying in Ann Arbor and does not often come home.

At the end of the 1940s, Goudsmit begins to give talks about his time in Nazi Germany and his pursuit of Hitler's nuclear scientists. Just as in his book *Alsos*, his conclusion is invariably that German bigotry and terror chased many important physicists away or silenced them. Loyalty to the Nazi regime was paramount, not scientific quality. As time goes on, however, his addresses begin to move more toward the present. In 1950, on a Tuesday evening in February, he speaks to the Methodist men's club in his hometown, Sayville. After the customary account of the German nuclear failure, the discussion turns to the new enemy, the Russians. Is it possible to reach an agreement about nuclear weapons with them, the audience wants to know. He really does not have any idea, Sam says in his ever-amiable way. "I don't know the Russians, I don't know the people, nor their land, nor their leaders. But I think that is at once also the core of our problem with the Russians: we don't know each other. We don't know them and they don't know us."*

According to the story in the local newspaper, however, he goes further: "Mr. Goudsmit added that we should never stop trying to prevent a new war. That is one nobody will win, he says. Maybe we will win on the battlefield, but it will leave us crushed and poor, and there will be nobody to help us." After the discussion, the traditional oyster soup was served, the reporter notes.

What the report says about Sam's work as head of the Physics Department at Brookhaven is noteworthy: "Goudsmit also says that the system with which the AEC judges the loyalty of its personnel is as fair as it can be." At the same time he criticizes the growing mistrust of Communists: "Actually it is a waste of time."*

In 1946, Congress passes a law establishing the Atomic Energy Commission. The AEC has to organize the application of atomic energy in the broadest sense of the word. The Commission is led by a nine-person advisory council, the General Advisory Committee (GAC). In December 1946, Isidor Rabi is chosen to be a member of the council, which is then chaired by the lawyer David Lilienthal. One of the advisory committees is led by the former scientific director of the Manhattan Project, J. Robert Oppenheimer.

The original intention is to move quickly toward civilian applications of nuclear energy. But matters take a different turn. In April and May of 1948, the Americans test a series of new A-bombs on atolls in the Pacific Ocean. The Russians answer on August 29, 1949, with their first nuclear test. It is the start of a nuclear weapons race that will persist into the 1980s. The Cold War has begun.

At the end of 1949, as soon as it is clear that the Soviets have a nuclear weapon of their own, a group in Washington begins to lobby for a still more powerful nuclear weapon, the so-called hydrogen bomb. The moving force behind that development is a wartime refugee physicist from Hungary, Edward Teller. The grouchy but brilliant Hungarian fiercely hates Communists and finds a powerful ally in Lewis Strauss of the AEC, a banker and retired admiral, and an intimate of President Harry Truman. Strauss pleads for a radical expansion of nuclear fire power, and he finds leading scientists such as Luis Alvarez, Ernest Lawrence, and, of course, Teller at his side. They all emphasize that a large-scale program must be founded to construct a hydrogen bomb as quickly as possible.

Opposing noises come from Oppenheimer's General Advisory Committee. There the discussion focuses above all on the question of whether a hydrogen bomb, which in theory can easily be a thousand times more powerful than an ordinary atom bomb, actually makes military sense. In one of his analyses, Oppenheimer says there are no military targets large enough to justify it. There are also major technical doubts. According to the ideas of that moment, a hydrogen bomb is much too big to be transported quickly, "except maybe with an ox cart," as Oppenheimer skeptically comments. When Teller and his colleague Stanisław Ulam devise a much more compact design for the hydrogen bomb, Oppenheimer admiringly describes it as "technologically sweet." But he maintains his objection, which is in fact moral: the H-bomb is too big to be used militarily; the weapon can do nothing except destroy whole societies. The H-bomb will not make the world safer. The contrary is the case: once weapons exist, history shows, they will always be used.

At the same time, Oppenheimer and his colleagues on the AEC's Advisory Council are far from being pacifists. Early in the discussion, they have recognized the Soviet menace and the necessity of deterrence. Instead of an impractical superweapon, the United States would do better to develop an extensive arsenal of tactical nuclear weapons: not bigger than everything the Russians have, but rather smaller and usable on a battlefield. The H-bomb should not get the priority, but rather such tactical nuclear weapons.

The protest has no effect. On January 31, 1950, President Truman signs the order to build a hydrogen bomb. If the Russians can build something like that, as advisers like Teller and Strauss claim, then of course he has no choice: "Then it has to happen." An enormous technological and scientific effort gets going under the leadership of Edward Teller and Stanisław Ulam among others. They work in Los Alamos and later get their own new lab in California, at Livermore near Berkeley. On November 1, 1952, the first test of a hydrogen bomb, code-named *Mike*, takes place on Enewetak Atoll in the Pacific Ocean.

For a few moments, a colorfully segmented bubble of light, miles wide, towers over the coral island. A stupendous fire ball rises high into the atmosphere, surrounded by rings of clouds. On the spot where the island was there remains only a crater more than a mile wide. The Cold War has reached a new stage.

After this, work on nuclear weapons, atom bombs as well as hydrogen bombs, takes up all the Atomic Energy Commission's time for years. The AEC's work is therefore systematically surrounded by military secrecy, and the scientists and technicians who are involved are all screened for possible subversive activities. Sam realizes early that such secrecy is on uneasy terms with the freedom of thought and speech that good science needs in order to get good results. He speaks regularly about that issue with colleagues he knows from the Alsos era. Into the 1950s, this Scientists' Committee on Loyalty Problems will meet regularly and discuss academic freedom. Goudsmit will continue to plead for a sensible and open approach.

"I doubt if all the scientists who have been turned down constituted a real security risk. Do memberships in left-wing organizations in the 1930s mean anything now, or that someone's wife organized a demonstration against the Spanish insurgents? Is it not much more important that someone has always been discreet, trustworthy and upright?" Of course, blackmail is a security risk, he writes matter-of-factly. "But if a scientist does things that his wife is not supposed to know about, for example, it is more efficient to tell his wife than to dismiss him. Then, whatever else may happen, the danger of blackmail is averted at once."

Security consciousness in the United States reaches obsessive heights after the German-born physicist Klaus Fuchs, a leftist, confesses that, while working on the Manhattan Project in Los Alamos, he passed American atomic secrets to the Russians. The presumption at that moment is that the first atom bomb constructed by the Soviets was based on that information. It is tested in 1949 and resembles a rough copy of Fat Man, the very first plutonium bomb constructed at Los Alamos in 1944–45. During the interrogation, Fuchs states that the married couple Julius and Ethel Rosenberg are implicated in the case. They are sentenced to death in 1951.

Tried in a British court, Fuchs gets a fourteen-year sentence, but in 1959 he is freed and leaves for East Germany. The Rosenbergs are executed in 1953, although later research has suggested that they may not have committed serious acts of treason. They leave two young children. One of the fiercest prosecutors in the Rosenberg case is later Republican President Richard Nixon, then still an ambitious young lawyer.

The Fuchs affair also raises questions about the wartime security of the bomb project in Los Alamos. Was sufficient care taken there with security and secrecy? And were all the hundreds of scientists and technicians working there equally loyal to the United States of America? Although the project resided under the general leadership of General Leslie Groves of the US Army, it is the scientific leader of the project, the theoretical physicist J. Robert Oppenheimer, member of the Advisory Council of the AEC, who comes into the spotlight.

After the war Oppenheimer has, in the public perception, become *the* face of the Manhattan Project and a modern American hero. Oppenheimer led a large team of brilliant scientists, men such as Enrico Fermi, Hans Bethe, and the very young Richard Feynman. He is generally seen as the real father of the atom bomb: a thoughtful, pipe-smoking, pencil-thin man with a razor-sharp intellect. In wartime photos he invariably stands, wearing a porkpie hat, a jacket and necktie, staring at the desert. Simultaneously determined and dreamy. In some photos General Groves stands beside him, in uniform shirt and cap, the belt pulled high around a substantial paunch.

In spite of all romantic images, American atomic secrets were spilled to the Russians under the nose of this brilliant Oppenheimer, or so is the sentiment around 1950. Where did things go wrong, is the question posed ever more loudly by politicians and the media.

They went wrong with Oppenheimer himself, anonymous sources are whispering around this time. This scion of a wealthy New York family is said to have moved in left-wing and even Communist circles during the 1930s, and also to have supported them financially. Who knows where his sympathies lie now? Oppenheimer, after the war one of the president's most important advisors in the nuclear realm, opts for the attack. In 1948, he gives an interview to *Time Magazine* in which he speaks openly about that controversial past. Yes, he says, he was a leftist, with even more leftist friends. But who was not? A youthful indiscretion he calls it, a life experience that has shaped him. "Until 1939 I was a fellow traveler, and after that the course of events meant that the travel rapidly declined." It is a closed chapter, Oppenheimer declares.

All the same, extensive FBI investigation brings incidents into view that make life tough for Oppenheimer. A question dealing with a professor of French, Haakon Chevalier, and his wife, Barbara, both employed by the University of California in Berkeley, is central. They are intimate friends of Robert and his wife, Kitty.

One evening, probably in 1942, the Chevaliers are visiting the Oppenheimers. Robert goes to the kitchen to mix Martinis. Haakon follows him and starts a conversation. He has just had a remarkable meeting with a good acquaintance, George C. Eltenton, a British physicist who works for Shell Oil and who is also acquainted with Oppenheimer. Eltenton asked Chevalier whether perhaps he can arrange for Oppenheimer to pass information about his classified work to the Soviet consulate in San Francisco.

Oppenheimer reacts as if stung by a wasp. He tells Chevalier that something like that would be treason, pure and simple, and that he wants nothing to do with it. He mixes the Martinis and goes back into the living room, where Kitty and Barbara are waiting with a medieval French book about plants that soon engages all four of them. The men soon forget the strange and uncomfortable conversation in the kitchen.

Until 1943, that is, when Robert is asked during a security check whether he has ever been approached by foreign agents. Yes, he says, and names Chevalier and Eltenton. Especially the British employee of Shell could be a problem, he says casually.

The authorities are alarmed at once. The next day, on August 26, 1943, Oppenheimer is grilled again, this time by an FBI officer named Boris T. Pash. Oppenheimer does not know the exchange is being recorded. And more than ten years later it is this, of all recordings, that plays a decisive role in hearings about his own reliability.

Oppenheimer tells Pash that Eltenton tried three times, via an intermediary, to link people in the Manhattan Project with a Soviet contact who knew what to do with microfilm. That intermediary was his friend Chevalier, often referred to in the documents as *X*. More than ten years later, however, Oppenheimer states in a hearing that he invented the whole story with the three attempts and the microfilm. Why, his questioner wants to know, and Oppenheimer shrugs his shoulders. "I made up a cock-and-bull story . . . because I was an idiot." Looking back, Oppenheimer says he would have been mainly worried that the story he told to the authorities was too vague, while being aware that there certainly might have been a security leak.

Oppenheimer's admission, in 1954, that he told a lie during a 1943 FBI interrogation, seems innocent, but it cooks his goose. Investigating counsel seizes on it and is able to sow increasing doubt on Oppenheimer's ability to manage affairs and supervise them, back in the day but certainly now, as chief advisor to the president in nuclear matters.

One of the decisive declarations in the hearings is that of the theoretical physicist Edward Teller, the man of the hydrogen bomb. Teller, a stocky man with a coarse face and beetle brows, was already working on the idea of nuclear fusion within a nuclear weapon during his time with the Manhattan Project in Los Alamos. There he had a reputation as a hawk when it came to both Nazis and Communists, and he has always had an aversion for Oppenheimer's arrogance and intellectual agility. According to many observers, professional jealousy plays a part. The contrast is between the quicksilver dandy Oppenheimer and the oaf Teller. He says in so many words to the AEC clerk, as they prepare for the hearing together, that he wants to make Oppenheimer look like a fool in his own domain.

Teller's testimony before the Gray Commission is one of the low points of the Cold War. On April 28, 1954, after almost three weeks of witness hearings, he takes his place on the witness stand in Washington and makes a declaration, assisted by special counsel Roger Robb. As on every other day, Robert Oppenheimer sits on a brown couch placed against the back wall in a long, rectangular hall full of chairs. The account of the gathering reads like a classical drama.

Is it Teller's intention, asks AEC counsel Robb, a tough trial attorney in daily life, to state that Dr. Oppenheimer is disloyal to the United States?

Teller: "I do not want to suggest anything of the kind. I know Oppenheimer as an intellectually most alert and a very complicated person, and I think it would be presumptuous and wrong on my part if I were to try in any way to analyze his motives. But I have always assumed, and I now assume, that he is loyal to the United States."

Robb: "Now a question which is the corollary of that. Do you or do you not believe that Dr. Oppenheimer is a security risk?"

Teller: "In a great number of cases, I have seen Dr. Oppenheimer act—I understand that Dr. Oppenheimer acted—in a way that for me was exceedingly hard to understand. I thoroughly disagreed with him on numerous issues and his actions frankly appeared to me confused and complicated. To this extent I feel that I would like to see the vital interests of this country in hands which I understand better, and therefore trust more."

And he adds sometime later: "If it is a question of wisdom and judgment, as demonstrated by actions since 1945, then I would say one would be wiser not to grant clearance."

According to eyewitnesses, Teller turns around after his cross-examination and, walking past Oppenheimer, who was sitting on the couch at the back, he extends his hand, and says: "I'm sorry." Oppenheimer shakes Teller's hand and replies: "After what you've just said, I don't know what you mean."

The hearings are completed on May 6, 1954, and the commission begins its work on a final report. It is issued more than two weeks later, on May 23. The decision, by two votes to one, is not to extend Oppenheimer's AEC security clearance. True, Oppenheimer is a loyal citizen but also a security risk, Chairman Gordon Gray says. In an explanation, the commission lists four main considerations: Oppenheimer's continuing conduct showed a disregard for security matters. He showed susceptibility to possibly risky influences. His conduct in the hydrogen bomb project is disturbing enough to raise doubts whether his continued participation would serve the best interests of the country. And Oppenheimer has been "less than candid" in some of his testimony.

"Their reasoning was tortured," two leading Oppenheimer biographers, Kai Bird and Martin Sherman, state in 2005, after all documents about the affair have finally been made public. The report consists of almost a thousand closely printed pages, totaling 750,000 words. "They did not accuse Oppenheimer of violating any laws or even security regulations. But his associations gave evidence of a certain indefinable ill-judgment." The reason for all this is Oppenheimer's years-long sharp opposition within the AEC. He has too often argued against the over-hasty development of the hydrogen bomb, Strauss and Teller's showpiece.

Right after the cancellation of Oppenheimer's security clearance, Goudsmit prepares a statement that will be published in the May 1954 issue of the *Bulletin of the Atomic Scientists*, but will also appear in other places. In his inimitable fashion, Sam puts into words the bewilderment and disbelief of many scientists over what has happened to their colleague: "The suspension of a scientist of such rare and great abilities as Oppenheimer possesses weakens our scientific strength and plays directly into the hands of our enemies," the declaration states. "No one who has followed his career can doubt Oppenheimer's devotion and complete loyalty to our country. Therefore I am convinced that he will soon be reinstated. Oppenheimer's case will then have a favorable effect upon the judgment of similar cases of lesser prominence. It should teach the public that wisdom can only be attained by broad experience and that a diverse opinion does not mean disloyalty. Oppenheimer's past makes him a better judge of true loyalty."

Then Sam mounts his hobby horse: "Remember that the German scientists lost the atom race because of insistence upon conformity of opinion. The superiority of democracy stems from its inherent consideration of varied and opposite views. Only opinionless nincompoops, yes-men and cowards never disagree and they also never contribute to progress. Oppenheimer's and some of his colleagues' original advice against speeding up the H-bomb program shows merely that this was a most difficult question even for the best-informed experts, and only those blinded by hysterical fear can see in it a sign of disloyalty or unfitness." Sam assumes that a favorable resolution of the Oppenheimer affair "will eventually silence hysteria and will help create again an atmosphere conducive to further strength and greater progress through scientific achievements."

Goudsmit and many others with him are too optimistic. Oppenheimer was not relieved of his function because he genuinely represented a danger to national security. The affair, almost all the historians say in the decades that follow, turned on control of the American nuclear arsenal.

After the report and Oppenheimer's definitive suspension, a nationwide protest gets underway among physicists and other scientists. A protest letter to Strauss is signed by 282 former Los Alamos physicists. A petition goes around the country, ultimately bearing the signatures of 1,100 scientists. A razor-sharp analysis appears in *Harper's* dismissing the Oppenheimer affair as a shocking judicial error. The only problem with this is that the suspension is not the result of a court case. That the AEC does not want to consult Oppenheimer any longer is, in fact, his problem and no one else's. Strauss does not see this as a matter of national interest.

In the following years, Goudsmit will regularly refer to the Oppenheimer case in letters and articles. His view is simple: the leftist past that Oppenheimer was confronted with in 1954, as well as his leftist friends, had long been known to the authorities and had never been a problem before. Even the Manhattan Project's General Groves said so to the AEC commission: Oppenheimer was loyal and effective as no other. However, this was unable to save Oppenheimer.

After his barbed comments about Oppenheimer's trustworthiness before the Gray Commission, Teller is persona non grata to most physicists. He is no longer invited to gatherings of professional colleagues. And when he does show up somewhere, people quite literally turn away from him. Years later, when he attends the presentation of a presidential medal to Oppenheimer by John F. Kennedy, everyone is surprised that Teller and "Oppie" do shake hands quite cordially.

It is no different for AEC chairman Lewis Strauss. In 1959, the proposal to invite him to the annual conference of the American Physical Society (APS) leads to a small uprising among the members; many see in the retired admiral and AEC leader the evil genius behind Oppenheimer's downfall. A petition goes around from the so-called Last Strauss Committee, which demands that Strauss *not* speak at the annual meeting. "The invitation gives the impression that Mr. Strauss enjoys the support and respect of the scientific community," it states. On the request of his friend and APS president George Uhlenbeck, Sam tries to pour oil on troubled waters. "We should not blow this up out of all proportion for political reasons," he writes. He does believe that Admiral Lewis Strauss is a key figure in nuclear issues in the United States.

On other occasions in those years, Sam Goudsmit acts in a conspicuously neutral way. In letters written to Edward Teller before and after the Oppenheimer affair, hardly any difference in tone can be discerned; Sam is generally quite businesslike. Usually the letters deal with the appointment of young physicists and only rarely touch on the personal. If Teller is in the neighborhood, he really should come by Brookhaven, Sam often writes cordially: "I would like to speak with you." But he does continue to wrestle with the Oppenheimer issue. In 1955, Teller tries, in vain, to publish a piece about the history of the hydrogen bomb in *Science*. While writing it, he asks Sam for advice a couple of times. Sam does not respond with complete sincerity. At this location in the Goudsmit archives, there is a slip of paper in Sam's handwriting, consisting of two words an exclamation point: "*Zum Kotzen!*" (Absolutely sickening!).

What Sam's politics are in those years is completely clear: diametrically opposed to hawks like Teller. Ever since President Franklin Delano Roosevelt, Sam has been a confirmed Democrat. For years, Goudsmit has also belonged to a number of socially critical organizations and committees, spawned by the fallout from the Manhattan Project, and he is one of the early advisors of the *Bulletin of Atomic Scientists*.

He makes no secret of his views about peace and nuclear weapons: according to Goudsmit, these weapons are too dangerous to be left to the military and have to be brought under international control via diplomatic action. But at the same time, he is working at Brookhaven National Laboratory, where nuclear research is carried out under the auspices of the same AEC that organizes nuclear weapon research with labs in Los Alamos and Berkeley (Livermore). Sam's involvement with AEC is not really technical in nature. He has never worked on nuclear weapons, and as a theoretician he is also not really the man for technical applications of nuclear fission. On the other hand, Goudsmit turns out to be an excellent source of information about the physics community in the United States. Like few others, he knows who is who, where the talented people are, and what their work is worth. He is called on regularly when security services want to check out physicists' backgrounds. As Cold War paranoia in the United States increases, with senators like Wisconsin's Joseph McCarthy hunting for everything that is red or looks that way, Sam's role proves more difficult.

Goudsmit knows one of the members of the AEC particularly well. Robert F. Bacher is just three years younger than Goudsmit. In 1930, he was Sam's first doctoral student in Ann Arbor to graduate, with a dissertation about the Zeeman effect. Later, they would together publish a much-used standard text about atomic spectra that also serves as a monument to Goudsmit's legendary intuition for patterns and links in the atomic tracings.

In 1940, they were once again working together at MIT, where the United States was doing everything to perfect radar, invented by the British some years before. From the start of 1942, Bacher was Oppenheimer's right-hand man in Los Alamos. As the scientific head of the Alsos mission in Europe, Sam wrote to Bacher regularly with questions and personal matters. The two have an extremely close friendship that will last until Sam's death in 1978. Bacher is, as a physicist, the only scientist in the AEC, which also consists of lawyers, bankers, and businessmen, every single one of them appointed by President Harry Truman.

When Oppenheimer becomes the topic of discussion, Goudsmit does not doubt for a moment which side he is on. For years he has been arguing for openness in a world that threatens to succumb to suspicion and paranoia. Because of the Oppenheimer affair, not only does security paranoia win out over openness, but scientists will think twice before they express their opinions openly. For Goudsmit, it is an unpleasant echo of what happened to German science in the 1930s under the Nazis. Through the repression of dissidents and critics, scientific thought not only became impossible, but also suspect and therefore dangerous. This led to an exodus of talent and propelled a generation of second-rate scientists into control.

Goudsmit has been peripherally involved in the Oppenheimer affair, as well. At a conference in 1950, he is approached by a fellow physicist and former Manhattan Project man, Bernard Peters. Peters asks Sam whether he can take a look at Peters's AEC security file. Peters is suspected of harboring Communist sympathies, something that is dangerous at that time. Peters says he was particularly shocked by something Oppenheimer said during the war, namely that he could tell, just by looking at Peters, that he could not be trusted. Sam examines Peters's file and can really cannot find anything objectionable in it. He, too, does not understand how Oppenheimer could ever have cast suspicion on Peters so thoughtlessly.

In discussions about nuclear arms, nuclear energy, and scientific responsibility, Goudsmit often comes into contact with Linus Pauling, one of the leaders of the antinuclear peace movement in the United States. In 1954, Pauling receives a Nobel Prize for Chemistry, and in 1962, another one, this time for Peace. Since the dropping of the A-bombs on Japan, he has been an outspoken and fierce opponent of all weapons of mass destruction. An early participant in the Pugwash Movement, he has worked assiduously toward the achievement of a test-ban treaty. Top scientist Pauling is not afraid to carry banners while leading peace demonstrations.

Goudsmit is cast in a more moderate mold: less outspoken, more pragmatic, more law-abiding, and above all, more nimble-footed. He does have his personal and political opinions, but he is also employed in one of the national laboratories in the nuclear field. That requires a certain reticence in the public domain.

The two men have known each other since long before the war, since their student days, in fact. In 1930, the book *The Structure of Line Spectra* appears, a thick standard work about the theory of atomic light emissions. It contains analyses of countless spectra, which betray, as though they were fingerprints, the identity of the atoms in question. Authors are the Dutch physicist Samuel Abraham Goudsmit and the brilliant young American chemist Linus Pauling, whom Sam met in 1927 at Niels Bohr's institute in Copenhagen. There, Pauling learned the fine points of quantum physics at the feet of stars such as Bohr and Arnold Sommerfeld. Goudsmit wrote his dissertation in Copenhagen in a few weeks, rushing to meet the deadline imposed by his imminent departure for America. Before he left, he had to get his doctorate.

Sam's dissertation, about line spectra of atoms, is written in Dutch. But when Pauling gets his hands on it in 1929, he decides, unasked, to begin a translation into English. That is a somewhat curious step, given that the American does not actually know the Dutch language. That turns out to be no hindrance. With his knowledge of German, a language close to Dutch, and equipped with a dictionary, the highly intelligent Pauling begins to translate. And when he is finished, he sends his work, which has gradually become an adaptation, to Sam in Ann Arbor. Pauling and Goudsmit are both listed as authors on the cover page.

Among colleagues, Pauling's cheeky maneuver calls forth amazement. Sam makes light of it. The important thing is that his work can now be distributed widely. For a long time, *The Structure of Line Spectra* will remain a standard book for nuclear physics, even when the new quantum theory of Heisenberg and Sommerfeld begins to set the tone.

In 1977, Goudsmit once more gets fully involved in heated discussions about nuclear issues. At that moment, public discussion focuses especially on the safety of nuclear reactors, also fed by fear of radioactive precipitation after decades of nuclear tests. Goudsmit believes the excitement about fallout to be a diversion that draws attention away from the key issue: nuclear arms themselves. "I find the endless debate about nuclear energy boring and irrelevant," he writes in the March 1977 issue of the *Bulletin of the Atomic Scientists*. "The chance of mass destruction by an atomic war is several orders of magnitude larger than that of the imagined accidents which might occur with nuclear plants." Moreover, he finds the risks that are being calculated for nuclear reactors to be far-fetched and pointless. The ironist in him states the matter thus: "One absolute danger to life is life itself; it is sure to end in death. There are no zero risks in human endeavors." In the same issue, Linus Pauling fulminates against the so-called need for nuclear energy.

Sam's past keeps pursuing him, but sometimes in a positive way. In June 1969, he receives an unexpected letter in Upton, written in a conspicuously curly woman's hand. It is from one Shirley R. Linfield in Evanston, Illinois, home of Northwestern University, where Sam taught for two years before his move to Brookhaven. Linfield has worked for the past few years as a technician in the Faculty of Science and in the course of cleaning up a basement office has come across something special: a box full of old photos. "Nobody had ever paid attention to it, even though it stood on a shelf in full view." In several photos, she recognizes Enrico Fermi, the famous physicist and Nobelist, and she first thinks the box must have belonged to him. But then she also finds photos that, according to the inscriptions on the back, were taken in Brussels and the Netherlands. There are images of families, travels, conferences, Niagara Falls, North American native people, Rome, Ann Arbor. And there is a photo of a girl, dated August 1932, with the inscription on the back: "*für Tante Goudsmit,*" for Aunt Goudsmit.

Goudsmit? Linfield's husband, a pharmacist at Northwestern, remembers Goudsmit and knows he is currently at Brookhaven. Hence the letter asking whether she should send the photos to him: "I am asking first, to prevent them going missing for a second time.""

Sam is flabbergasted. His photos! A week after the letter from Evanston arrives, he sends a reply, handwritten, though that is not his usual practice. Of course he is happy, and of course he would like to have them back after all those years. But first he takes the time to say how the collection was lost in the first place. That he left Michigan in 1941 for war work, first at MIT and later in Europe. That he stored his belongings in Ann Arbor, but never went back after the war. That when he was writing his 1947 war book *Also* (Sam calls it "a book about my hilarious adventures in Nazi Germany""), he needed his old papers and wrote to Ann Arbor whether they could send him his old stuff. That a few boxes did arrive, but that they turned out to contain nothing but offprints of old articles. The rest seemed to have gone missing. "They assumed that in the general tidying mania after the war, a janitor in Michigan had thrown out all my things with other stuff. I even got apologies from the chairman of the physics department.""

Years later, a colleague finds a box of books with Sam Goudsmit's name in them in an Evanston basement. When this colleague brings them along to Brookhaven in 1957, one box turns out also to contain letters, photos, and negatives. It includes letters to and from fellow physicists from 1921 to 1941. "It was a wonder,"" Sam says.

The discovery of a lot more photos is a new wonder, one whose historical value Sam recognizes, although in the well-known almost-apologetic Goudsmit tone. "Informal snapshots of physicists of that period [1930s] are quite rare, most of the photos were posed. But I messed around a bit with my cheap hand-held camera with f 2.8 and a shutter speed of 1/10. There was no fast film yet, so most of the photos are underlit and badly focused. I once used a couple for a historical article, which I am enclosing."'

Today Sam's photos are stored in the Emilio Segrè Visual Archives of the American Institute of Physics in College Park, Maryland. The archives hold many hundreds of his photos, from the carefree summer schools in Ann Harbor in the pre-war years to the hunt for Hitler's nuclear experiments during the last year of the war. Sam is regularly depicted, often in uniform and generally with a happy or excited smile.

In the 1950s, Sam Goudsmit seems to be having the time of his life in many ways. In photos from those years he looks relaxed, often with a cigarette and always in a light jacket with a white shirt and bow tie. He smiles a lot, even when he is not looking into the camera.

In the spring of 1958, he plunges himself into a new adventure as an editor that will definitively establish his name as a periodicals man. The April 15 issue of *Physical Review*, the flagship journal of the American Physical Society, carries a brief announcement of a new competing title, *Physical Review Letters (PRL)*. Sam explains why a new scientific journal is badly needed and why the rules of publishing will have to change. On July 1, 1958, the first issue of *PRL* appears; it will be published every two weeks. In the years that follow, the journal full of technical scientific discussions grows quickly into the most important medium worldwide for physicists. With respect to speed and impact, it is comparable only with the British journal *Nature*, which is authoritative especially in the biological field. Sam remains editor-in-chief of *PRL* until he formally retires in 1970.

The idea behind *PRL* is simple: it is actually the letters section of the much slower monthly *Physical Review*. With new methods of typesetting and offset printing, it becomes possible to publish letters within two to three weeks of arrival, instead of the customary six to ten weeks. In addition, the new journal publishes summaries of longer publications that will later appear in *Physical Review*. Such things, Goudsmit writes in his first comment, should improve communication among physicists, speed up the stream of ideas, supply information more quickly, and prevent the duplication of labor.

Offset has disadvantages as well. It is much faster than traditional typesetting, but intervention becomes much more difficult. Things almost go wrong with the first issue. One author (Sam does not mention a name) wants to make a change after submitting a piece but does so too late. The only thing that remains is to withdraw the whole piece by pulling out all the stops, accompanied by costly overtime and delay. This can never be allowed to happen again, Goudsmit warns in the second issue. The third issue carries a letter that was messed up so badly by the new technology in the first issue that it is not possible to correct it by publishing a list of errors.

Only genuinely new results qualify for *PRL*: discoveries or pieces about hot topics in the discipline. The rules are rigorous. Moreover, the letters have to be accessible to non-specialists: the physics must be presented clearly. At the same time, the pieces are not allowed to be longer than two pages and cannot include more than two figures.

Furthermore, there is an immediate warning: speed has a cost in quality. Goudsmit: "Since speedy publication allows no time for thorough refereeing, the Editor is likely to make mistakes and to include occasionally Letters of minor importance or below our usual standards. Such occurrences cannot be used as a precedent to require the Editor to accept similar letters later on." For the same reason, manuscripts have to be publication-ready, checked, and supplied with the correct references. "We have no time to do library research." There is not much time for real peer review, Goudsmit also says; many decisions will be made by the editorial staff. George L. Trigg will be the assistant editor, reporting to Goudsmit.

Physical Review Letters is an experiment in publication, Goudsmit writes in the first issue on July 1, 1958. If things go wrong, there is emphatically a way back to the "Letters to the Editor" column in *Physical Review* and a slower tempo of publication. "We intend to make changes and improvements in the course of time." Things do go wrong, but in a different way than Sam expects. The problem actually is that *PRL* is an immediate success. The journal introduces an unprecedented tempo into the world of scholarship, at a time when everything still has to circulate in paper form. From the start, the logistics begin to pinch.

The almost casual style of Goudsmit's editorials is striking. In the September 1958 issue, he explains in detail how the type system used by the editors works. For the first time, strange symbols and signs that are not found on the keyboard of a standard typewriter can be typed all the same. In the first issue of 1959, Sam not only wishes his readers and authors a happy new year but also a year that will bring no more new particles and fewer but better theories. "We shall do our best to keep *Physical Review Letters* thin enough so that our readers can learn about all the latest significant developments in physics research without having to wade through bulky volumes." Meanwhile, "we believe the subscription rate to be ridiculously low . . . for the service rendered."

In a different way, Sam is also a founding father of a novelty in scientific publication: in 1960, he invents the press embargo. Since *PRL* has become the fastest-publishing scientific journal in the world, next to *Nature*, inventions published in it are often of interest to a larger public. That is nice, but in Goudsmit's view, it is also a problem. To be of service to the media, he would like to signal them about potential news stories. But to this he adds the demand that nothing be published until the relevant issue of *PRL* appears. In his opinion, authors must make the same demand when they contact the press, he writes in the issue of January 1, 1960: "Scientific discoveries are not the proper subject for newspaper scoops, and all media of mass communication should have equal opportunity for simultaneous access to the information. In the future, we may reject papers whose main contents have been published previously in the daily press.... Careful planning and preparation of news material will achieve increased accuracy in the interpretation of science."

Newspapers, Sam thinks at that time, too often show a preference for idiots with wild theories. Surely a serious scientist does not want to be confused with them. And Sam can be tough. Many years later, two astronomers, William E. Howard III and Lewis E. Snyder, recall a 1969 incident when they submitted an article of which the press had gotten wind. From the October 2009 issue of *Physics Today*: "... The paper progressed smoothly through the PRL review process. Everything blew up around 5:30 one evening, when editor Goudsmit, en route home from the West Coast, read in the *New York Times* that the ... results had been leaked and picked up on the news wires. We spent about 30 minutes on the receiving end of a Goudsmit telephone barrage in which he said over and over that because of the newspaper leak there was no way that *PRL* could publish the article.... At one point during the barrage, we made the mistake of noting that [George] Trigg, who was much more understanding, had already locked the article in the *PRL* press. Goudsmit roared: 'In that case, I'll jerk the article out and send through blank pages!' His anger eventually subsided after he realized we had nothing to do with the *Times* leak, and he went ahead with publication."

The astronomers learn a few weeks later that a "professor from a prominent northeastern university had heard about the discovery and had leaked the news to a newspaper reporter." One of them also heard from someone who was in the *PRL* office at the time of the exchange, "... that when Goudsmit slammed down the phone, he looked up, grinned, and said something like: 'That'll hold 'em. Let's go get coffee!'"

Sometimes he seems to deal with the power of the editor as if he is playing a game. A NASA researcher is speaking with Goudsmit at a party, and they are discussing the journal's reputation for rigor (it rejects a third of all submitted work): "I told him that nevertheless *PRL* had never rejected a piece of mine. He immediately got out his pen and an envelope to write on and asked me jokingly: 'What's your name again?'"·

In the 1960s, a revolution takes place in scientific publishing, and Sam occasionally struggles with the rapid developments. To the extent that the personal interests of researchers increase, it becomes more important to ascertain who has been the first to discover something. In an editorial in the September 27, 1965 issue of *PRL*, Goudsmit pokes fun at a proposal to set up a register of preprints, initially in theoretical physics: "The next step might be to equip theorists with portable recorders so that all of their statements about physics, including those uttered in their sleep, would be preserved on tape. The contents of the tapes would be transmitted electronically to interested colleagues via a distribution center." It is all a joke, of course, but Sam does worry about scholarly publication, which seems to be getting completely out of hand because of the growing quantity of authors and research. "... It is obvious that the centuries-old system of communication by journal publication is no longer adequate however much it expands, and that significant changes are needed. The nature of the changes is not yet known. Perhaps they will be as radical as the invention of the alphabet in its time. The presently fashionable computer retrieval methods are probably only a partial answer. It will soon be possible to retrieve old material so rapidly and exhaustively that it may overwhelm the user as completely as do the present bulky journal issues."

On October 21, 1974, Sam writes his very last editorial, giving it the title "Swan Song." "According to the Constitution and Bylaws of the American Physical Society I will have reached statutory senility on the day of the next Council meeting, 25 October. This is the end of my career as Editor. On this occasion I am expected to make a few profound statements about changes which occurred in physics over the years. I am also expected to predict the future, but I refuse to do that in print. I claim that physicists have not changed, but physics has.

"Physics research has grown in size, cost, and, in the eyes of outsiders, also in importance. Physics was, of course, always the most important subject to active physicists, but before 1940 the public, the press, the government, and the military rated it far below stamp collecting."

Nowadays, Goudsmit continues, physics has become a matter of "short-lived fads and fashions," a few of which stick in the public mind while the names that pertain to them are soon forgotten. At the same time, the physicist is considered to be an entrepreneur, while in Sam's view he is at heart a kind of artist in his creative urge.

One of Sam's standing jokes in speeches about his journals *Physical Review* and *Physical Review Letters* in the later years is that at the current rate of growth, by 2000 *PRL* would be growing faster than the speed of light. "That is a matter of simple extrapolation. That this does not contradict the theory of relativity is because information transfer will have sunk to zero by that time." This is the kind of humor physicists like.

According to Sam, the situation leads to problems for the journals and their editors: "... The editorial office is plagued with complaints about priorities, biased reviewers, failures to cite previous work, and sometimes even plagiarism. ... Some of the skills I have acquired on this job might be useful in managing an opera company, where the difficult prima donnas are at least prettier than most of those with whom I have had to deal up to now. ... I have also learned to write friendly answers to unfriendly letters." This is Sam Goudsmit to the core: with a self-possessed cheerfulness that always puts the sharp things he unavoidably says into perspective.

Sam's vision for *PRL*, former editor Daniel Kleppner writes in the February 2009 issue of *Physics Today* on the occasion of the *PRL*'s golden anniversary, "combined an elevated view of physics with a somewhat tempered view of physicists. Reading Sam's editorials today, one hardly knows whether to laugh aloud or weep that his world of physics, in which one could reasonably aspire to be broadly knowledgeable and follow important advances in every field, has vanished, blown away by the explosive growth of physics and the physics community, the increasing complexity of research, and the revolutionary impact of digital communications."

In the 1970s, Goudsmit, jovial in company, always smoking a cigarette, invariably in white shirt and bow tie, is more and more considered to be a grand old man of American physics. Virtually all physicists know him, if only because of his central position at *Physical Review Letters*. That is the most sought-after physics journal in the world: current, fast, always full of the newest insights and results.

As an old-timer, Sam also regularly involves himself in more political issues and gradually begins to sound ever more radical. Without being contradicted, he writes in incisive letters, this government will slide into becoming a totalitarian regime that will not differ much from that of the Communists in Russia. In 1970, he travels with a group of scholars to Washington to make an appeal to the administration to end the war in Vietnam. Their position is that the war is robbing America of an entire generation of talent and diverts money and attention from the country's real problems: poverty, race issues, and environmental pollution.

The scholars are received in Washington by Lee DuBridge, an old acquaintance of Sam's who, during the war, ran the radar lab at MIT where Sam worked for a while. "Now Lee was scientific advisor to President Nixon, and we tried to make it clear to him that the war in Vietnam is wrong," Sam says around this time in an interview with a Dutch newspaper, the *Haagse Post*. In it, he confesses that he has ever more trouble in seeing the importance of science, and of physics in particular. "I grew up before physics had anything to do with defense. Science was something good, then, like music or a painting. Just like a Rembrandt or a Mondrian, science was something valuable. Now I am no longer so sure of that. At the same time physics and mathematics are all I can do. If that is taken away from me I am a nobody." This is the gloominess that will increasingly characterize Sam Goudsmit during his final years.

In November 1973, Sam writes a long piece in his office at Brookhaven, Upton, Long Island, that reads like his intellectual will and testament. It appears in the February 1974 issue of the *Bulletin of the Atomic Scientists*, and subsequently also in other journals and newspapers. "I Am Scared," is the title.

The piece begins with a remark made thirty years ago, when an American officer explained to him that the Americans would undoubtedly use their atom bomb: "That was about six months before our first test and eight months before we dropped A-bombs on Hiroshima and Nagasaki.

"Since that time the USSR, France, China, and the United States have continued to stockpile A-bombs with the excuse that this acts as a deterrent against war. But for how long? There are always minorities and generals willing to use them. Moreover, we may soon have an all-volunteer army of well-paid professional soldiers, which could be a first step towards a government takeover by a military junta."

At the same time, the issue of nuclear disarmament seems to have lost its appeal. Instead, the protesters have been diverted into concern about fallout and pollution. Very worthwhile objectives, but totally irrelevant when compared to the threat of nuclear war. "In September 1959, I warned in the *Bulletin* against this diversion of effort, writing: 'I strongly recommend stopping scaring the public about fallout: this fear will not help abolish war, but may hamper progress in the beneficial uses of radiation.' ... The emphasis on safety is delaying construction of nuclear reactors. This aggravates the energy shortage which could lead to an old-fashioned war for oil sources, claiming many more victims than the hypothetical effects of reactor efflux. If the safety addicts and the computer model prophets had been around when Henry Ford worked on his Tin-Lizzy, they could have foreseen the hundreds of thousands killed or crippled every year by such contraptions and stopped its development. We would now all be going to work on recyclable bicycles." Sam, born in the Netherlands, thought this would be all right, at least in the summer.

He continues: "Talking about pollution, I want to point out that the worst offenders are not engineers and physicists but members of the medical profession. They pollute the world with a population excess, the basic cause of pollution and shortages. ... They also keep old people, like myself, going on for many years beyond retirement, after they have stopped all productive and constructive activities. This creates . . . a big drain on the economy, although it is negligible compared to military spending. The narrow-minded and shortsighted foes of birth control don't see that massive violence is the only alternative for regulating the natural population limit. Moreover, wars are no longer fought merely on the battlefield; they envelop all of society . . ." Sam gets criticism as well as support from older colleagues. More than twenty newspapers in the United States reprint the piece. Goudsmit's "I Am Scared" is a hit.

CHAPTER 9

UNDER THE SPELL
OF THE SCARAB

Leiden–Nevada, 1925–1978

On a hot summer afternoon in August 2015, a small group of interested persons who are gathered in the lobby of the stately Kelsey Museum for Archaeology in Ann Arbor, Michigan, raise their glasses to the exhibition that has just opened: "Passionate Curiosities." The title is a deliberate play on words: the exhibition is about curiosities in private collections of arts and artifacts, as well as about the curious spirits who have brought them together. Samuel Abraham Goudsmit, former professor of physics in Ann Arbor, among other places, is one of them.

Esther Goudsmit, Sam's only child, is absent from the opening, the curator of the exhibition, Margaret Cool Root, notes. Now and then, usually in the city's post office, she runs across the aged Professor Emeritus Goudsmit, a frail woman whose intellect is still sharp. Ms. Goudsmit is not in the best of health, observes Root, who with her conspicuous long platinum-white hair is herself no longer in her first youth. Esther Goudsmit is old and tired. But what can you expect when you are eighty, Root muses, as she contemplates the visitors at the opening.

The exhibition in the university museum will run until the end of November, and in the months after the opening, the exhibition catalogue turns out to be a big success. The stories by the archaeological collectors, in combination with their often quite remarkable private collections, draw many visitors and many book buyers. And one of the collections, the Egyptian antiquities of the late Professor Sam Goudsmit, speaks particularly to the imagination, all the more because the Dutchman initially stumbled into Egyptology by mistake but ultimately acquired considerable authority in it. What is more, in the 1970s, he tried to tempt the National Academy of Sciences (NAS) into financing an X-ray machine especially for mummies and other ancient artifacts. That this proposal was declined appears from a short letter in the immense Goudsmit archive in College Park, Maryland. "Research of that kind lies far outside our usual interests," was the NAS's frugal response.

Sam's interest in Egyptian antiquities has its origins in his Leiden study years. He told the story many times, but one of the best versions can be found in a fine profile of Sam that appeared in two parts in the *New Yorker* for November 1953. A profile in this well-known weekly is no mean honor for a professor of physics, no matter how exciting his life story. The author of the articles is the journalist Daniel Lang.

Sam's interest in classical Egypt is inadvertent, but happenstance leads him into a world of magic, beauty, and mystery hitherto unknown to him. It all begins in the Huygens Society, which Sam Goudsmit joins as a student around 1920. The group meets weekly, and it is the customary for members to take turns in presenting scholarly papers. Sam, too, speaks occasionally, but in time there are complaints about his rather limited choice of subjects.

Goudsmit invariably speaks about atomic spectra, a subject that engages him almost obsessively and that in many respects is the reason he began to study physics in the first place. As an eleven-year-old, he was captivated by the idea that lines in the light of the stars betrayed the same chemical elements as existed on earth. That, at least, was the claim of a physics text that his older sister left lying around the house on Koninginnegracht in The Hague.

But the other members of the club are starting to get enough of Sam's atomic tales. The president entreats the enthusiastic young physicist to devote his next turn as speaker to a completely different subject. It does not matter what, so long as it is not about spectra. Goudsmit is taken aback; he has the sense that he really knows nothing about anything else. But in order not to disappoint the president, he registers in a course in Egyptology. It should be possible to find a good subject there, he thinks.

When Sam opens the classroom door for his first semester, he turns out to be the only student to have registered in the course. The professor is Pieter Aart Boeser, the leading Dutch Egyptologist, who takes the matter lightly. Three participants are enough for a seminar, Boeser says in adapting a Latin proverb: God, the teacher, and one student. And so Goudsmit reluctantly starts out on the study of Egyptology. For two years Sam does not want to disappoint Boeser, a small, timid man with an imposing mustache and gentle manners, and he attends the private seminar weekly, even when he does not really have time.

He has little trouble learning to read hieroglyphs, but when the instructor wants his colleague Ehrenfest's young student to immerse himself more deeply in the subject, Goudsmit drops out. Instead of reading texts, he comes under the spell of scarabs, the small, sacred Egyptian figures in the form of beetles, made of jade or other stone. "The professor considered that vulgar, but the reality of the scarabs made me feel closer to Egypt than the printed hieroglyphics I'd been working on."

While he is still studying with Ehrenfest in Leiden, in the early 1920s Sam is also a part-time assistant in Nobelist Pieter Zeeman's laboratory in Amsterdam. Now and then he has time to go into the city center. At those times he likes to frequent the art galleries in the Spiegelkwartier, the area tucked between Herengracht and the Rijksmuseum. The Amsterdam art trade is at that time a world center for the visual arts and antiques. Especially renowned is the supply of art from the ancient world. At one of the dealers, D. Komter, Goudsmit finds what he has been eagerly seeking for some time, an antique Egyptian scarab. For Komter, Egypt is peripheral; mostly he buys and sells old paintings. From his small collection, the self-assured Leiden student picks a scarab in excellent condition and pays ten guilders for it. Sam has a bit of money these days; he earns 400 guilders annually in Zeeman's lab.

During the transaction, Komter quickly realizes that he is faced with an enthusiast who really knows the score. He suggests to Goudsmit that he take home Komter's entire collection of 120 unsorted scarabs and classify them. Sam is pleased and goes to work in his room in The Hague. In the weeks after the meeting in Amsterdam, he dates, photographs, and catalogues all the pieces. With his elementary knowledge of hieroglyphs, he translates the inscriptions. Komter gratefully accepts the fruit of Sam's labor and as a reward gives him a few small, beautiful figurines, which he later always identifies as the real start of his collection of Egyptian art. Not long after the event, Komter closes his store. He is retiring. At an auction in Amsterdam, the entire collection of scarabs is sold for 120 guilders. Sam knows about it, but on his low salary he cannot afford to buy them.

Many years later, as he himself is nearing retirement, Goudsmit looks back with amusement on his first steps in Egyptology in a Summer 1972 article for the archaeological journal *Expedition*. Much has changed in Amsterdam's Spiegelkwartier since the 1920s. Antiques are no longer curiosities and have become genuine *objets d'art*. "What you dig up out of the ground is no good for the art trade" is a doctrine he ascribes to a famous Dutch-English family of art dealers. "It was not until the Second World War that the . . . doctrine of art versus archaeology did an about-face. Ancient Egyptian curios became 'Art', and . . . the prices have skyrocketed—which takes me out of the competition, although my income is now somewhat more than $400 [*sic*] a year." Because money has come to play an important role, the matter of distinguishing genuine from fake is now crucial.

"Laymen often believe that there are such things as experts who can judge authenticity of an antique merely by looking at it. The eye of the expert can indeed detect qualities that escape the layman. There is nothing supernatural about this. We recognize the handwriting of friends. . . . When a friend writes in haste, or when he is ill, his handwriting changes, but usually we still recognize it. . . . Similarly, an art expert can recognize the style of a painter and be absolutely sure when a signature does not fit a painting."

However, evident mistakes and ugly details in art objects are not necessarily evidence that there has been funny business. There is a simple explanation: "Books and art museums acquaint us with only the very best workmanship. However, in the Cairo Museum and in the tombs it is easy to find numerous examples of lower grade workmanship. . . . I have seen forgeries with more appeal for the art connoisseur. . . . What you dig up out of the ground is now very good for the art trade. But it is perhaps forgivable for a longtime Egyptology buff to view this reversal of values with a doubtful eye and a twinge of regret."

At the same time that Sam and Jaantje arrive in Ann Arbor, in 1927, plans are developed in the Department of Archaeology, led by Francis Kelsey, to open the university's own museum. The Kelsey group carries out a lot of field work in Egypt, in a large and yield-rich excavation at Karanis, near El-Faiyum. The Americans excavate there until 1935 and countless pieces are carried off to Ann Arbor.

Sam finds this interesting and talks regularly with Kelsey and other Egyptologists about his own Egyptological studies. Professor Kelsey determines that, for an amateur, the Dutch physicist reads hieroglyphics with astonishing ease. Sam's Egyptian collection will ultimately comprise dozens of pieces, from scarabs and seals to amulets, papyrus fragments, figurines, burial gifts, jewels, and textiles. In a number of instances, these are fragments of larger earthenware objects or papyrus rolls, as extensive research reveals. Some seem to have been broken or torn deliberately. Sam suspects that shrewd Egyptians dealers do this themselves in order to get more money for more objects.

One of the most beautiful pieces in the Goudsmit collection in the Kelsey Museum is a slab of limestone on which a profile of a woman's face is drawn in ink. It is clearly a sketch, a rough draft. On it two additional, fainter profiles can be discerned, plus a lion's head and an arm.

All are sketches for motifs on a larger monument. Goudsmit bought it in 1959 at a public auction of excess holdings of the Metropolitan Museum of Art in New York. Evidently, the museum had no more need for it. He was particularly happy with it because it reveals a rare informal moment in Egyptian antiquity: an artist who first attempts something before he decides how to proceed.

Egypt continues to appeal. Sam even has an Egyptian student when he briefly teaches at Harvard during the early war years. The student sees an opportunity. He claims that back home in Cairo he can put in a good word for his American instructor and arrange a temporary teaching appointment for him. In exchange—I'll scratch your back if you scratch mine—the student wants to get passing grades on his exams. He gets those, but in the end, he does not return to Egypt at all, an amused Goudsmit later recounts: "The rascal married an American girl and started a family in Boston."

Even during the war, Egyptology briefly plays a role in Sam's life. In August 1945, the Alsos mission reaches Berlin. It is a few days before the A-bomb attacks on Japan, as a result of which Goudsmit will be flown back to Heidelberg in a headlong rush. The Allies want to prevent the Russians, at that time the masters of Berlin, from seizing Sam as a nuclear scientist.

Before that happens, Goudsmit has halted his jeep at the great Archaeological Museum on Unter den Linden, which has been heavily damaged. He knows about the imposing Egyptian collections that German explorers and travelers have carried to it, and he asks himself with disquiet what can be left of it all. In his own German period, as a student in the 1920s, he has stood by the display cases filled with papyrus rolls, mummies, reliefs, and statues. Now he finds only an exhausted German security guard surrounded by enormous chaos in the soot-blackened building. The man recognizes a connoisseur in the American visitor, and he tries to give Sam an entire mummy case. When Sam refuses, the guard breaks three pieces of painted papier-mâché from the ornamented lid and presses them into his hands.

They are images of Isis and her sister, Nephthys, kneeling to the left and right of Osiris, the god of the underworld, plus a depiction of a falcon with spread wings. Sam estimates that both are from 525 BCE. Dumbfounded, he puts the fragments in the pockets of his army great-coat. Ultimately, he will take them along to America, where they can still be seen in the Kelsey in Ann Arbor. According to the experts, Isis's kneeling pose means mourning, intended to bring the deceased Osiris back to life in the eternal cycle of death and resurrection.

The same pose may be seen in a wooden Isis figurine that Sam buys in 1941 from a London art dealer, Spink & Sons, while he is in England for secret work on radar systems for the Royal Air Force. The figurine is from the once famous collection of the banker Frederick Hilton Price. It was found in northern Egypt, but never really dated. Years later, Sam gives it to his daughter, Esther, who donates it to the Kelsey Museum in 1981.

Egypt will continue to engage Sam all his life. Shortly after the war, he writes a proposal for a large-scale physics project to measure cosmic radiation, in which the pyramids of Gizeh play an important part. Goudsmit proposes that Geiger counters be placed inside and outside the stone colossuses in order to measure how penetrating the radiation from the cosmos potentially is. The project does not proceed, and it will ultimately not be until the 1960s that Goudsmit actually does travel to Egypt a couple of times. At that time, the United States has a strong diplomatic relationship with the North African country. On several occasions, Goudsmit tries to be helpful to people in the struggling country after he has traveled in the Middle East in 1964 and 1967, travels he also uses to breathe in the culture and the antiquities of the region.

In Cairo—he cannot help himself—he buys a number of antique pieces from dealers. At the same time, he writes in reports that he is shocked by the sloppy, rather unprofessional manner in which, for example, a museum like Cairo's National Museum of Antiquities manages its riches. Nowhere is there decent equipment for analyses, and the primary goal of restoration is to make things look beautiful again, a mortal sin among archaeologists.

A few times, the amateur archaeologist Sam Goudsmit shows his true nature. Three times he writes a scholarly article that is actually published. Each time, the subject is objects in his own collection. The articles are negligible compared with the hundreds of publications in physics that Goudsmit has written in the course of his long career. Nevertheless, they have a special place in his heart. Over the years, he regularly writes about them to his friends and colleagues, always in much the same language: "A nice piece of work, if I say so myself. And although it is just one page, I am quite proud of it."

In 1976, he sends an addition to his entry in *Who's Who in America*. "To make it a bit more lively, you could add: 'Graduated from Leiden University, where he studied Egyptology as a minor.'" The letter is Goudsmit to the core: "From my point of view none of these additions are essential. What I never really like about this sort of biography is that it looks as if I have not done a thing in physics since 1925!"

In 1962, Sam gives almost his entire collection to the Kelsey Museum as a permanent loan. That happens almost in passing, in a letter to the director about a completely different subject. After his death in 1978, his wife donates the collection definitively to the Kelsey, exactly as Sam has arranged it in his will.

According to an inventory prepared by Margaret Cool Root for an exhibition in 1982, the Goudsmit collection comprises scarabs, seals, amulets, illustrated papyrus fragments with hieroglyphs and texts, demotic and Coptic pot shards, stones with inscriptions and reliefs, fragments of decorated mummy covers, figurines made of wood, bronze, terra cotta and stone, pottery funerary dolls, jewels, and Coptic textiles. There is a unique limestone slab with unusually informal sketches. A seated statue from the New Era. A prehistoric painted pitcher. A dish in the form of a diving girl for the grinding of cosmetics. An alabaster mortar. Lids for pitchers. The wrappings of a mummy from the grave of Horemheb. In total, twelve crates with objects go to the museum; together they contain more than 124 large and small objects.

A fine example of Goudsmit's characteristic outlook on the art that fascinated him so much is a study of an old papyrus fragment that he published in the January 1974 issue of the *American Journal of Archaeology*. "An Illiterate Scribe" is the title of the piece, comprising only two columns, but it shows how here, too, Sam's intellectual capacity proves useful in discerning patterns, so important in the atomic spectra he once studied.

In antiquity, texts on papyrus were frequently copied. The fragment that Sam is examining, a depiction from Am-Tuat, the funerary book he bought in Europe, is, he knows, a copy of the much older original that is kept in the Louvre. For unknown reasons, it is executed in a secret script, although in a simple manner: the texts do not run from right to left, as is customary in Egypt, but the other way around. The copyist does not seem to know that and copies the hieroglyphs from right to left on a smaller piece of papyrus. Therefore, he has letters left over and puts them on the right side of the next line instead of left. As a result, the copy differs from the original in Paris in a way that proves the copyist could not read what he wrote. "Such systematic errors cannot be explained as accidental oversights or carelessness; they are proof of the illiteracy of the copyist." In this way, Goudsmit's renowned feeling for regularity in complicated patterns, well-known among his colleagues in nuclear physics, brings a long discussion among Egyptologists to a conclusion: ancient texts were copied not only by scholars but also by humble, uneducated manual workers.

Sam's good friend and colleague George Uhlenbeck calls him "a magician in dealing with cryptograms. He is perhaps not circumspect, but he had and has an amazing talent for taking random data and giving them meaning." Isidor Isaac Rabi of Columbia University, and chairman of the Atomic Energy Commission, knows this characteristic all too well: "Sam has a sixth sense for finding order in a chaos of facts. He thinks like a detective; no, he is a detective."

This is accurate. During his time as research assistant for Zeeman in Amsterdam, Sam hears about a chemist working for the local criminal investigation department who is giving a course on techniques of investigation. Goudsmit registers in it and in the course of eight months learns everything about fingerprints, ultraviolet light, and the analysis of blood stains and fat. "Taking that course was one of the most sensible things I ever did," he says years later to an interviewer for the *New Yorker*, Daniel Lang. "I learned a detective's attitude toward the evaluation of evidence. It's been extremely helpful in my work. . . . People usually go too much by intuition, without judging what's in front of them."

But physics remains front and center. From September 6 to 11, 1965, Dutch physicists are in Amsterdam to celebrate the one hundredth anniversary of the birth of Nobelist Pieter Zeeman, the legendary discoverer of the magnetic division of spectral lines. The company of hundreds of Dutch and foreign physicists attend several days of an intensive lecture program in the Royal Tropics Institute, at the top of Plantage Middenlaan and not far from Zeeman's old lab on Plantage Muidergracht. Sam Goudsmit is there as well, but he earlier politely turned down an invitation to give an overview of the Zeeman effect. He really does not understand it all that well, he says with his eternal tendency to self-deprecation: "More has happened in a half century than I realize." Sam uses the gathering above all to meet old acquaintances. He has no part in the official program.

In the 1960s and 1970s, Goudsmit regularly visits Europe and the Netherlands. He usually gives a few lectures and chiefly visits old friends and family. Occasionally, his Dutch friends seek his advice on issues of research policy. Around 1950, he has given advice on the establishment of a new central research organization in the Netherlands, the organization for pure scientific research ZWO, later renamed NWO, the Netherlands Organisation for Scientific Research. He is also, in the background, one of the driving forces behind the establishment of the National Institute for Subatomic Physics (Nikhef) in Amsterdam.

And sometimes he is once more fully his own obsessive self. In the 1960s, Goudsmit undertakes a special search for the personal archive of one of his old mentors, Pieter Zeeman.

In 1963, Sam takes up contact with one of the heirs, daughter Jacqueline Zeeman. She writes Goudsmit, who was once her father's lab assistant, that there was discord in her family after her mother's death a year earlier. She and her younger brother have left the parental home on Stadhouderskade after a quarrel with her older brother, who now lives there alone. Her father's letters and documents are probably somewhere in the house, but she does not know where. And to go looking for them is impossible, given the disagreement. With that, the matter seems to have reached a dead end.

But Goudsmit does not throw in the towel. In 1971, he calls on a number of Amsterdam physicists for help, among them the later Professor Pieter Klinkenberg (not related to the *Waarheid* journalist) of the Zeeman lab on Plantage Muidergracht. Klinkenberg turns out to be in contact with Jan Zeeman, the obstinate son living on Stadhouderskade. He says he is going to see how the land lies. But it is doubtful whether the prized letters are actually there: "Zeeman did not use a typewriter, so there are no carbon copies of letters. Probably it makes more sense to recover addresses and to inquire there." What follows is a long list of potential recipients, which ultimately Goudsmit cannot do much with.

Only at the end of the 1990s, after the death of Jan Zeeman, does the Amsterdam historian of science Anne Kox find boxes full of material in an upstairs room of the house. He secures the letters and documents before the municipal waste services department cleans out the totally run-down and garbage-filled house. At last, the Zeeman archive is in place. By this time, Goudsmit has been dead for years.

In part because of the Cellastic affair centered on Professor Jaap Kistemaker of the University of Amsterdam, the Dutch media lie in wait for Goudsmit when he visits the Netherlands. He is interviewed several times by the weeklies *Haagse Post* and *Vrij Nederland* among others. "The mischievous humorist among the physicists," the *HP* reporters Michel Korzec and Bert Vuijsje characterize him as in 1971. "More than once that playful tendency has come back to haunt him, because he did not sense that a lot of other people took themselves very seriously indeed."

The conversation takes place in the lobby of an Amsterdam hotel, some days after Goudsmit said during a press conference that Jaap Kistemaker had taken his medicine by now, and that Cellastic had nothing to do with atom bombs. He tells Korzec and Vuijsje that he was shocked by what the newspapers wrote about this during the days that followed, led by press service ANP and the daily newspaper *De Telegraaf*: that Goudsmit had judged Kistemaker very harshly. In reality, he has had it with the unceasing attention paid to the war, especially in the media. That the war and the death of his parents have always hung over his life like a shadow he regards as a private matter about which he rarely speaks.

Characteristic of that reticence is a letter he writes in 1976 to the sometime military leader of the Alsos mission, Colonel Boris Pash. In August, Pash walks unannounced into the physics department of the University of Nevada, Reno, where Sam teaches part-time in retirement. By chance Sam happens to be there, but he is just back from a trip to Europe, including the Netherlands, with Irene. He is battling jet lag. He has come only to pick up his mail and has no more than a brief conversation with Pash. A few days later, he apologizes for the somewhat curt exchange in a friendly letter, in which he sounds disillusioned more than anything else.

People, he writes, are no longer interested in stories about the Second World War. "Even historians are fed up with it. If you talk about *the* war, they do not know exactly which one you mean, Vietnam, Korea, Sinai, or maybe even the First World War, of which it was said that it was the war to end all wars." Even the men of the former Alsos team seem to have put history behind themselves. "They rarely look back on our glorious adventure of thirty years ago. Some even say they would simply like to forget it. That's because they are not so happy about the outcome of the war. I myself find it difficult to accept that our former enemies, Germany and Japan, are at this moment beating us economically. Their competition in the area of cars, electronic goods and photography is costing us jobs here. How can that be? It is because in those countries it is prohibited by law to invest tax money in armaments. They put it into industrial development. In reality they have won the war."*

In the interviews reported in the Dutch media in the early 1970s, Goudsmit generally minimizes his big breakthrough in physics: the discovery of the spinning electron in 1925, when Sam was twenty-three. "People such as Heisenberg, Pauli and Kramers had reached the conclusion that the concepts of classical physics never offered a solution to the atom. For them the electron was a point of view, almost philosophical. But I was young and did not know much about physics, and I was no genius, not by a long shot. As a consequence I kept imagining an electron as a little ball that can turn on its axis. That was good enough for an interim model of the atom and an explanation of the spectra. That discovery was in large measure a fluke. Of course it turned out later that Pauli with his abstract approach was much more nearly right."

Sam and Uhlenbeck's discovery of the electron spin in 1925 brought them international fame and jobs in the United States. Without that breakthrough, Goudsmit says in the interviews, he would probably have become a teacher deep in the backwoods. "Maybe in a secondary school in Middelharnis [a small town in the islands southwest of Rotterdam]. That was proverbially the only vacancy at that time."

Goudsmit considers the attention paid to him to be rather exaggerated. He thinks mythologizing is at work in the United States, too, for that matter: "People in America always think it is such a big deal, for example, that I knew Einstein personally. In the 1920s he came to

Leiden for a month every year, visiting Ehrenfest. There were six or eight of us students there, so you could not possibly avoid Einstein, even if you had wanted to. Now people think... My God, it is as if you dated Marlene Dietrich or something. Back then it was all so unimportant!"

And the times were different in science itself, Goudsmit emphasizes: "In the 1920s physics was rather like collecting stamps. If you made a discovery in spectroscopy you felt as if you had found a mistake in an overprint or something. An uncanceled specimen. Nothing more. Not until just before the war did physics suddenly become important for armaments. The physicists were the only ones to understand anything about electronics and because of that we were able to develop radar. We were the only ones who understood anything about neutrons, and because of that we were able to come up with the atom bomb."

In Germany, things went very differently in the run-up to the war, Goudsmit says. "The German physicists were unable to get the Nazi government interested to the same degree, perhaps because they did not believe very strongly in it themselves. As Albert Speer writes in his autobiography: 'They did not ask for a lot of money, and people who ask for so little cannot be good.'"

Besides, Europe and America were different worlds in the 1920s. When he arrived in America in 1927, Goudsmit noticed this at once. "Every American student was alive to potential technical applications. All of them had taken apart an old Ford at some point and had reassembled it. For that reason it is no accident that a machine like the cyclotron was invented in America. Only people who knew that large magnets exist could do that."

At the end of 1976, Hendrik B. Casimir, president of the Royal Netherlands Academy of Arts and Sciences, and director and director of research at Philips NatLab, writes a letter to Sam Goudsmit and George Uhlenbeck. He has a request. On July 7, 1977, it will be fifty years since Goudsmit and Uhlenbeck received their doctorates at Leiden. "Your Dutch colleagues do not want to let this noteworthy anniversary pass without notice. Would you two be willing to come to the Netherlands for this golden occasion?" Casimir invites the two guests of honor to a two-day colloquium, including festive social events. Travel and accommodation expenses will be paid, for their spouses as well. Casimir wishes the two a happy 1977 and makes excuses for what he calls his "rather amateurish typing," used as he is to secretaries.

George immediately lets Casimir know that for family reasons he will not be available during the summer. Sam does accept the invitation and begins working on a speech about the discovery of the electron spin, the phenomenon with which he and George earned world renown. He shows the text to Uhlenbeck for additions and comments. But Uhlenbeck is unexpectedly critical. He does not much like the matter-of-fact way in which Goudsmit describes the discovery of spin as a fluke. "You deprecate yourself too much," he comments on a passage in which Sam says that he could not follow Lorentz's initial skepticism and criticism at all. "In my opinion it is also too strong to describe all your work about quantum regularities in atomic spectra as 'numerology' and to say it was not theoretical physics. It is *so*, because the half-guessing at the regularities always (just as with elemental particles) has to be done first, before the correct synthesis is found."

Uhlenbeck says he *is* touched by Sam's personal tone. "One can really hear your voice, it is just exactly the way you are: deprecating yourself too much, very generous in regard to others and especially me, skeptical and disillusioned about yourself, and perhaps about science as well, alas."

Sam pays attention to his friend's criticisms and amends several passages. His criticism of historians of science becomes somewhat milder. He speaks with appreciation about, among others, Thomas Kuhn (who once interviewed him at length about his work) and the German historian Martin Klein, who has done a lot of work on Heisenberg. The word "numerology" has disappeared, it has become "formalism . . . or another milder way of putting it."

The 1977 speech has been lost, but it must have been an adaptation of the story that Sam had already told in 1971 at the fiftieth anniversary meeting of the Netherlands Physics Society in Leiden. The discovery of the electron spin, he said there in some many words, was only after the fact seen as a major contribution to science. "But I did not know that yet, they only told me afterwards. You do not have to be a genius to make a big contribution to physics."

Goudsmit modesty was wasted on his colleagues. The Columbia University physicist and Nobelist Isidor Isaac Rabi: "Physics must forever be in debt to those two men for discovering the spin. . . . Why they never received the Nobel Prize will always be a mystery to me." In 1964, the two get a Max Planck Medal for the discovery, awarded to them on the urging of, among others, Werner Heisenberg. In 1976, the two also receive the National Medal of Science, one of the highest distinctions for scientists in the United States.

The great appreciation for Goudsmit's work is most evident in the archives of the Nobel Prize committee in Stockholm. The nominations for the annual science prizes are kept secret for half a century, chiefly not to cause embarrassment to anyone, neither the supporters nor those being applauded. From the documents that are now open it turns out that in the course of the years Sam Goudsmit was nominated no fewer than thirty-nine times for the Nobel Prize in Physics. On almost all occasions, he was nominated in combination with his friend and colleague George Uhlenbeck, invariably for the discovery of the electron spin.

Between 1947 and 1964, the twosome was regularly nominated by a number of prominent colleagues, among whom Léon Brillouin, Otto Stern, Glenn Seaborg, Victor Weisskopf, and John van Vleck are best-known. Seaborg, the discoverer of plutonium, who himself received a Nobel Prize for his work, nominated the two Dutch-Americans on three occasions. The very first nomination came from the French physicist Léon Brillouin, who nominated Goudsmit as early as 1935, though without mentioning Uhlenbeck, which is strange. But instead of Uhlenbeck he nominated Wolfgang Pauli, the Austrian theorist who is chiefly known because of the exclusion principle: electrons never occur twice in the same location in an atom. With this rule, the light spectra of atoms such as hydrogen are easy to explain. *Why* that rule exists Pauli did not know, either, until Goudsmit and Uhlenbeck introduced their spin, an extra characteristic that means that apparently identical electrons can still differ.

Abraham Pais, another Dutchman who emigrated to the United States and had a career there as a theoretical physicist and a historian of science, discusses the matter of the electron spin a number of times, in his books about Einstein and Bohr, among others. The issue turns out to be more complicated than it seems to be.

Goudsmit and Uhlenbeck, Pais writes, did not know in 1925 that their German colleague Ralph Kronig already had the idea of electron spin in January 1925, when he heard about Paul's exclusion principle while visiting Tübingen. According to Kronig, he hit at once on the idea of an extra characteristic of the electron, a spin or whirling. "This afternoon I succeeded in deriving from it the so-called relativistic doublet formula," he wrote in his logbook. Wolfgang Pauli, the originator of the exclusion principle, had a very low opinion of Kronig's approach and also pointed out a mathematical error. Prominent people like Heisenberg and Hendrik Kramers also expressed their doubts, and ultimately Kronig decided not to publish.

The messy run-up to the discovery of the electron spin is, according to Pais, the real reason a Nobel Prize has never been awarded for the electron spin. It can even be maintained that the American Harold Urey and the South Asian (Indian) Jagadish Bose had similar ideas, too, but did not dare to publish them, mainly because of the advice of senior colleagues. After all, how can a point particle turn on its own axis? In Pais's opinion, the history of the electron spin teaches the lesson that young physicists should not listen to older professional colleagues: "The researcher's fate can also depend on the decision to heed authority, or instead to strike out on your own." Uhlenbeck has always been clear on this point; he says that Kronig did have the idea earlier but should have published it.

In the summer of 1970, Sam gets a letter from John Maddox, the editor of the famous British scientific journal *Nature*. Maddox asks whether Sam wants to write an article about the discovery of electron spin. The two have known each other well for years; Maddox has himself worked as a theoretical physicist. Moreover, he and Sam have, as editors, corresponded regularly about manuscripts that have been submitted to one of them and rejected and then published by the other.

Goudsmit is reluctant to commit himself to the idea of an overview article. "I regularly give lectures about the subject, but these are very informal and do not lend themselves to publication in a serious journal. Part of the success of the lectures comes from the slides I use and from my jokes, which won't come across on paper. I could write an article based on them, but it would be more suitable for *Punch* rather than a serious publication like *Nature*."

In any case, Sam has little self-confidence where the history of physics is concerned. He does not respond to a Dutch request to write about physics in the 1920s. "The reason is that I am not a historian and that, to be quite honest, I do not have the faintest idea what was happening in those years. . . . My Leiden story is limited. My major subject was spectral lines, my first minor was spectral lines and my second minor was spectral lines."*

Nature's John Maddox is not easily deterred. He proposes that Goudsmit should write a book in collaboration with George Uhlenbeck. That could be a monograph for a new series about the history of modern physics that he is organizing with a Danish publisher. That may be easier, he thinks. Sam reacts almost fearfully: "You evidently do not know either of us very well. George is a perfectionist who has been working for twenty years on another book that I suspect will be finished around the end of the century. Besides, he and I know nothing about spin. The developments since the 1930s have totally overwhelmed us. The idea of the electron spin and its modern consequences lie far outside George's interest and far outside my competence."* It hardly needs saying that the book about electron spin never appears.

In 1972, Sam Goudsmit retires from Brookhaven and the American Physical Society at age seventy, as professors are apt to do. But he does not really leave. He shows up at the lab regularly and will not in fact close the door for the last time until 1975. That year, Irene and he sell their house in Bayport, Long Island, and move to Reno, Nevada. The move is not illogical. For years Sam has been a board member of an organization that wants to build a research center for desert research, the Desert Research Institute (DRI), very near Reno. It is one of the subsidiary functions that he has taken on as part of his role as the central figure in American physics. He certainly likes the initiative. Energy self-sufficiency and the environment are important research projects, in his opinion.

More is going on after his official retirement in 1972, however. Financial worries have been running through Sam's mind all his life.

He has had prominent jobs and filled important positions. The Goudsmits have been able to live well, but he has never earned a lot of money. Besides, old family feelings play their part. It troubles him again that the businessmen in the Goudsmit family made fortunes and he as a scientist did not; certainly now that he can no longer go through life unworried, traveling, dining, and sleeping on expense accounts. This is one of the reasons he takes on additional projects for which his elegant writing style and his agreeable personality are his key qualifications.

He is no stranger to such extra paid employment. In 1966, he writes a popular scientific book about time for Time Life, with the science journalist Robert Claiborne. Aside from being a writer, Claiborne is a folk singer and an explicitly left-wing figure. As an ex-Communist, he experienced the Cold War personally. In 1960, he lost his job with *Scientific American* because of unpatriotic comments made during a humiliating hearing before the House Un-American Activities Committee (HUAC). Goudsmit, who is made of progressive, liberal material, gets along famously with him.

After his retirement, Sam receives a request from the Sloan Foundation to develop a physics course for non-physics students. Gradually it becomes clear to him how he should tackle this project: by teaching physics to medical students at the University of Nevada, Reno. In the process, he can become more closely involved in desert research.

In 1975, Sam and Irene find a ranch-style wooden house, surrounded by trees and lawns, in a housing development on Dickerson Road in Reno. Through the French doors opening from their modest living room they can hear the Truckee River gurgle by over the stones. The location is relatively quiet and cool, especially at the end of the day when the sun is low in the west. The campus is just over a mile away by car. The DRI lies farther outside the city, to the northwest, but it is also easy to reach.

The peace and quiet of Reno turns out to be most welcome. In a letter to a female friend, Sam explains how hectic the last years at the Physical Society were: "When I joined the publishing program in 1952, around 4,000 pages of scientific papers appeared annually, mostly in the *Physical Review*. Now there are 38,000, divided among five journals, of which the *Physical Review Letters* appears every week. Every month some 600 manuscripts arrive; there are thousands of scientific reviewers at home and abroad. *PRL* accepts about half of the submissions. There are constant goings-on between authors and reviewers. The editors are under permanent pressure from both sides; some time ago I had to change to an unlisted phone number at home. But that pressure is now in the past, and that is a great relief."*

In his letters of that time, Sam sounds happy about starting a new life in Reno: "Once we have been in Reno for a year or longer, we'll be nobodies. And that is what I have been hoping for. Our income will be minimal, and you won't find us in expensive restaurants. Irene is looking for work as a doctor's assistant. We love Reno and the people we have been meeting here since 1960. Therefore we expect that our move will prove to be a happy one and will lead to permanent settlement here."*

After the summer of 1975, Sam begins with his course, a low-key job that guarantees a modest but secure income. He has imposed strict conditions on the university. He will not sit on any committee, assume no administrative responsibilities, and will offer no advice on governance. He wants to be treated like an outside visitor. "Everyone who wants my advice can get it in the bar over a martini. But no more than that. My experience, though indeed substantial, is of little value under the present circumstances, I would guess. Personally I believe more in the wisdom of the young."*

The scholarly world does not let go of him easily. In 1977–78, for example, an impressive number of American scholars and scientists, led by Hans Bethe, one of the fathers of the American atom bomb, make an appeal for nuclear disarmament. Sam Goudsmit also signs this declaration, which is directed to President Jimmy Carter and Congress. "Signed and returned" he has written on the copy of the petition in his personal papers. The petition is clear. "The nuclear arms race . . . is increasingly a mortal threat to all humanity . . . More than ever it is urgent now to slow down and ultimately stop the nuclear arms race." Goudsmit has never been closer to pacifism. Among those drafting the petition were Linus Pauling and John Kenneth Galbraith, both vocal peace activists, each with a Nobel Prize on his resume. Politically, Goudsmit is closer to the center. But nuclear arms control, he thinks, is more important than partisan politics.

Goudsmit also continues to be a member of the advisory council of the American Physical Society, dealing with matters that have to do with the publication of their journals. On November 16, 1978, little more than two weeks before his death, a cheerful Sam shows up at a meeting of the publications committee of the APS in New York. That he seems to be his old self is not altogether as expected. In May of that year, Goudsmit became unwell at home, and a physician in Reno's hospital determined that he has heart problems. But with some rest they seemed to disappear.

In recent years, Sam has always been healthy in spite of his age. He weighs less than before and is a bit slenderer, but he suffers somewhat from painful joints as a result of gout. After the death of Irene's uncle in New York, Irene and Sam have not had a regular family doctor. Irene, a doctor's assistant, monitors Sam's blood pressure. It is 150/70 or 160/80. "Satisfactory," she says to her seventy-four-year-old spouse, who is still at the university on a daily basis. When Sam applies to the Ford Foundation for a traveling fellowship a bit later, he has to submit an official certificate of good health. The couple's friend Isadore Rosenfeld is a physician, and after a bit of prodding he issues that certificate: "Dr. Goudsmit is in excellent health and fit for travel and service overseas."" He does prescribe Benecid-500, a medication that suppresses chronic joint inflammation, a consequence of hyperuricemia.

The media still show up regularly at the Goudsmits' door during the 1970s. One of the last times this happens is when a Brazilian television crew want his commentary as an Egyptologist for a documentary about the supernatural power of the pyramids. Sam is not interested and writes a short tirade to explain why. The letter is Sam to the core: decided, funny, and sharp to the point of unpleasantness: "There is not one iota of trustworthy, reproducible and measurable proof that pyramids do anything, good or bad. Horseshoes and four-leaf clovers do not bring good luck. Friday the 13th is not necessarily a bad day. But people seem to want to believe more and more in idiotic fantasies such as the secret power of pyramids, flying saucers with extraterrestrial beings, life rhythms and other nonsense. The same thing was happening in Germany, just before Hitler came to power."" The war is never far from Sam's mind.

According to Goudsmit, the popularity of superstition shows that people have lost the moral and ethical guidelines of established religion. "At the same time, people have also lost their trust in the rationality of modern science, even though it has lengthened our lives and raised our living standards. Science and technology have brought us atom bombs and pollution, which overshadow the positive accomplishments. We are scared of the future; we sense that our societies are sick, very sick. The symptoms are religious and racial wars, overpopulation, and huge nuclear arsenals. We know that despairing sick people go in search of idiotic quacks. Our world is in that state. The belief in demagoguery and the supernatural means that we expect terrible things to happen, such as a nuclear war or epidemics caused by humankind. Of course a belief in irrational nonsense will not prevent them. What will prevent them I do not know."

This is not the first time that Sam gets worked up about enthusiasm for the supernatural. Many years before, in the early 1950s, he was a member of an official commission of inquiry formed to provide calm advice about the UFO hype that reached a high point at that time. The newspapers are full of reports of flying saucers, not only by excited citizens but also by military men and pilots. Right up to the presidential level, the authorities are pulling their hair out, especially because the Cold War is taking place and there might be Russian airplanes or rockets among all that flying stuff. The Air Force unit commanded by Major Edward Ruppelt charged with making an inventory of the UFO sightings is totally overwhelmed, although it is not clear why this should be so. To sort all this out, the Robertson Panel is established in 1953, named after its head, the physicist Howard P. Robertson. Among its members are sober-minded physicists like the later Nobelist Walter Alvarez and Samuel Goudsmit.

The plan is to do the work quickly and efficiently, if only to accommodate the crowded timetables of the scientists involved. The panel meets on a Wednesday and examines the evidence of sightings. There are photos, films, reports, and accounts, all sorts of things. The members of the commission offer a quick assessment and devise natural explanations for what may seem extraterrestrial. They finish writing their advisory note by Friday. Its conclusion is simple: a general nervousness is prevalent as a result of which countless normal phenomena in the sky are taken for flying saucers, from clouds and arcs of light to birds and airplanes. Briefly put, it makes no sense at all to deploy military means to resolve the UFO question, whatever public opinion makes of it and however the newspapers sensationalize it. Major Ruppelt's shop can be closed down.

Years later, Alvarez admits in a discussion with UFO believers that everything went just a bit too hastily in that 1953 panel, but that above all he was convinced that flying saucers and extraterrestrial life are a lot of baloney. "And I still think so." Goudsmit later expresses himself in similar terms: "In my opinion the subject of flying saucers is a total waste of time and money. The phenomenon should be researched by psychologists instead of physicists." It is a conservative point of view that will be held against him by UFO devotees for years afterward.

In the autumn of 1978, September or October, the British TV journalist and scientist Peter Goodchild visits the Goudsmits in their low-slung wooden house in Reno. The producer has become intrigued by the Alsos mission carried out by Sam and his colleagues during the last year of the war in Europe. Goodchild wants to make a long documentary about it. For hours he talks with Sam about his wartime adventures in England, France, and Germany. About his terrible visit to the emptied parental home in The Hague. About the insanity of SS headquarters in Berlin, where he found a child's skull in the fireplace. They talk about Werner Heisenberg, and about the fact that the Americans did construct a bomb and used it and the Germans did not. Sam takes Goodchild into the desert, to Pyramid Lake, a large lake with a remarkable tufa that rises out of the water in the form of a pyramid. The blistering heat, in combination with the water and the bright light, leads the two men to remain standing on the shore for a while, sunk in thought.

As Goodchild is leaving, Sam asks a question: might Peter be able to find copies of *Alsos*? He has just one himself, and the book has long been out of print. Back in England, Goodchild places a small advertisement and soon gets an offer of a copy from a book dealer. The book is not only in excellent condition, but the previous owner has put a clipping of a review from a 1948 issue of the *Times Literary Supplement* neatly folded in the back. Goodchild sends that copy and the clipping to Goudsmit on November 30. "Possibly you did not see the review at the time," the journalist writes. "Please accept the book and review as a token of my appreciation and as thanks for your hospitality during my visit."

Goudsmit will never see the letter of thanks and the old review. On December 7, 1978, Irene Goudsmit writes a four-paragraph letter to Goodchild that under the circumstances sounds remarkably calm and well-considered: "I must share with you the terribly sad news that your letter arrived one day too late. On December 4, Sam collapsed in the university parking lot with an apparent heart attack and died almost at once." Sam probably had not seen the review from the British periodical, she continues, and it would certainly have pleased him.

Irene writes that she is in any case happy to have the copy of *Alsos*. "Now we can lend it to friends without fear or hesitation, because it is no longer our only copy. We both enjoyed meeting you and the time you spent with us. I am so happy that you were in time for an interview with Sam, and that he was able to show you Pyramid Lake, a spot that meant a lot to him. Sam was a special person."

Gradually, the world learns what happened that Monday morning. Sam drove from his home on Dickerson Road to the university, a couple of miles away. He put the last touches to a physics exam for the medical students he had been teaching since the summer. He finished the problems, gave a few instructions, and then went outside, to his car. He sat down and must suddenly have lost consciousness. A random passer-by found him in the car, unconscious behind the steering wheel. He called for an ambulance, but it was already too late. Sam was dead on arrival at the hospital. A fatal heart attack, the duty physician concluded.

Watchful as always, Walter Sullivan writes a four-column obituary in the *New York Times* that appears on December 6. Science editor for the newspaper and considered to be the dean of science writers, Sullivan has no doubt about the most important event in Sam's life: "Samuel Goudsmit, Codiscoverer of Electron's Spin, Is Dead at 76." His work at Brookhaven and during the war are also mentioned at length, as well as his editorship of *Physical Review* and the trail-blazing *Physical Review Letters*. "Long a leading figure in American physics," is Sullivan's assessment.

At the beginning of 1979, the editors of *Physical Review Letters* write a final tribute: "We last saw him at a meeting of the Publications Committee of the American Physical Society held in New York on 16 November. Sam was then, as always, warm and witty, sardonic and wise, and above all, enthusiastic—enthusiastic about the prospects and the future of the publications he had shepherded so long and enthusiastic about the teaching he was doing at the University of Nevada. Possessing a piercing and critical intelligence, a warm personality, and energy and strength of character, Sam left behind many monuments: in science, especially in atomic physics, in government service during and after World War II, in administration through his chairmanship of the Brookhaven National Laboratory Physics Department . . . , and in publications through his management of the main journal of the American Physical Society, the *Physical Review*. Though few men have contributed more than Sam to the shape of the world physicists know today, the private memories many of us have of Sam's kindness and consideration, all salted with his bluff wit, are not less important than his more concrete achievements."

The evening after Sam's death, his wife, Irene, phones his daughter,

Esther, who lives in Rochester, Michigan, with her mother, Jaantje. It is already after midnight in the Midwest. Esther and her mother are startled by the ringing telephone. Irene reports that in his will Sam has left his body to science. Esther and Jaantje do not need to come to Reno, because he did not want a funeral or other memorial.

And that is how it happens, even though a number of friends and colleagues do organize a small ceremony for Sam at the university. Irene tapes the speeches, and she later sends a cassette to Esther. "First listen to it yourself, and then decide whether your mother can handle it or not," she writes in an accompanying letter. Esther does not play the cassette to Jaantje.

Jaantje Logher Goudsmit dies in Michigan on January 5, 1980, thirteen months after Samuel Abraham Goudsmit, the man with whom she had come from The Hague to America in 1927. She was young and elegant, he was full of scientific excitement, and both were full of expectation for their new life, far from The Hague and his Dutch middle-class background. It is all so very long ago.

BIBLIOGRAPHY

Baggott, Jim. *The Quantum Story: A History in 40 Minutes.* Oxford: Oxford
    University Press, 2011.
———. *Atomic: The First War of Physics and the Secret History of the Atomic Bomb.*
    London: Icon Books, 2015.
Baldwin, Melinda. *Making Nature: The History of a Scientific Journal.* Chicago:
    University of Chicago Press, 2015.
Ball, Philip. *Serving the Reich: The Struggle for the Soul of Physics Under Hitler.*
    London: The Bodley Head, 2013.
Bar-Zohar, Michel. *The Hunt for German Scientists 1944–60.* London: Arthur
    Barker, 1967.
Beevor, Antony. *D-Day: The Battle for Normandy.* New York: Penguin Books,
    2009.
Bederson, Benjamin. "Samuel Abraham Goudsmit (1902–1978)." *Physical Review
    Letters* 101 (1): 10002. Woodbury, NY: American Physical Society, 2008.
Bernstein, Jeremy. *Oppenheimer: Portrait of an Enigma.* Chicago: Ivan R. Dee,
    2004.
———. *Hitler's Uranium Club.* Woodbury, NY: AIP, 1996.
Bird, Kai, and Martin Sherwin. *American Prometheus: The Triumph and Tragedy of
    J. Robert Oppenheimer.* New York: Alfred A. Knopf, 2005.
Blacket, P. M. S. *The Military and Political Consequences of Atomic Energy.* London:
    Turnstile, 1948 (copy from Goudsmit collection).
Boterman, Frits. *Duitse daders: de jodenvervolging en de nazificatie van Nederland
    (1940–1945)* [in Dutch]. Amsterdam: Amsterdam Antwerpen Uitgeverij De
    Arbeiderspers, 2015.
Bouguerra, Mohamed Larbi. *Pauling* [in Dutch]. Amsterdam: Veen Magazines,
    2009.
Broad, William. *Teller's War: The Top-Secret Story Behind the Star Wars Deception.*
    New York: Simon & Schuster, 1992.
Cassidy, David. *Uncertainty: The Life and Science of Werner Heisenberg.* New York:
    W. H. Freeman, 1992.
Cohen, Ernst. *Ik had mij vast voorgenomen, niet over mijne toekomst na te denken* [in
    Dutch]. Ultrecht: Uitgeverij Matrijs, 2011.

BIBLIOGRAPHY

Conant, Jennet. *109 East Palace: Robert Oppenheimer and the Secret City of Los Alamos*. London: Simon & Schuster, 2005.

Crease, Robert, and Charles Mann. *The Second Creation: Makers of the Revolution in Twentieth Century Physics*. London: Quartet Books, 1996.

Delft, Dirk van. *Heike Kamerlingh Onnes: een biografie* [in Dutch]. Amsterdam: Bert Bakker, 2005.

Donderi, Don C. *UFOs, ETs and Alien Abductions*. Charlottesville, VA: Hampton Roads, 2013.

Eickhoff, Martijn. *In naam der wetenschap. P. J. W. Debye en zijn carrière in nazi-Duitsland* [in Dutch]. Amsterdam: Nederlands Instituut voor Oorlogs Documentatie, 2007.

Egleton, Clive. *The Alsos Mission*. Sutton, Surrey: Severn House, 1997.

Feynman, Michelle. *Perfectly Reasonable Deviations from the Beaten Track: The Letters of Richard P. Feynman*. New York: Basic Books, 2005.

Finkbeiner, Ann. *The Jasons: The Secret History of Science's Post-War Elite*. New York: Viking, 2006.

Fölsing, Albrecht. *Albert Einstein: Ein Biographie* [in German]. Berlin: Suhrkamp, 1995.

Goudsmit, S. A. *Alsos*. London: Sigma Books, 1951.

Groves, Leslie. *Now It Can Be Told*. New York: Da Capo, 1983.

Hawkins, Eliot. *Quantum Mechanics for the 99 Percent*. Amazon Digital, 2012.

Hoffmann, Dieter. *Operation Epsilon: Die Farm-Hall-Protokolle oder Die Angst der Alliierten Vor der deutschen Atombombe* [in German]. Reinbek: Rowohlt, 1993.

Irving, David. *The Virus House*. London: William Kimber, 1967.

Jones, R. V. *Most Secret War*. London: Wordsworth Editions, 1978.

Jong, Lou de. *Het Koninkrijk der Nederlanden in de Tweede Wereldoorlog*, part 10b 'Het laatste jaar' [in Dutch]. Amsterdam: Staatsuitgeverij, 1982.

Jungk, Robert. *Brighter Than a Thousand Suns: A Personal History of the Atomic Scientists*. Harmondsworth (England): Penguin Books, 1958.

Karlsch, Rainer, and Heiko Petermann, eds. *Für und Wider "Hitlers Bombe": Studien zur Atomforschung in Deutschland* [in German]. Münster: Waxmann, 2007.

Khalili, Jim al-. *Quantum: A Guide for the Perplexed*. London: Weidenfeld & Nicolson, 2003.

Klein, Martin J. *Paul Ehrenfest: The Making of a Theoretical Physicist*. Amsterdam and New York: Elsevier, 1970.

Lichtblau, Eric. *Nazi's in Amerika—Hoe Amerika een thuishaven bood aan Hitlers mannen* [in Dutch]. Amsterdam: Meulenhoff, 2015.

Maria, Michelangelo de. *Fermi: natuurkundige in een bewogen periode* [in Dutch]. Amsterdam: Veen Magazines, 2012.

Mehra, Jagdish. *The Golden Age of Theoretical Physics*. River Edge, NJ: World Scientific, 2001.

Monk, Ray. *Inside the Centre: The Life of J. Robert Oppenheimer*. London: Random House, 2012.

Orzel, Chad. *How to Teach Physics to Your Dog*. New York: Simon & Schuster, 2010.

Pais, Abraham, *Niels Bohr's Times: In Physics, Philosophy, and Polity*. Oxford: Oxford University Press, 1991.

———. *A Tale of Two Continents: A Physicist's Life in a Turbulent World*. Oxford: Oxford University Press, 1997.

———. *Inward Bound: Of Matter and Forces in the Physical World*. Oxford: Oxford University Press, 1980.

Pash, Boris T. *The Alsos Mission*. New York: Award Books, 2006 (original 1969).

Powers, Thomas. *Heisenberg's War: The Secret History of the German Bomb*. Cambridge, MA: Da Capo, 2000.

Rae, Alastair. *Quantum Physics: A Beginner's Guide*. Oxford: OneWorld, 2005.

Rhodes, Richard. *The Making of the Atomic Bomb*. New York: Simon & Schuster, 1986.

———. *Dark Sun: The Making of the Hydrogen Bomb*. London: Simon & Schuster, 1995.

Rispens, Sybe. *Einstein in Nederland: Een Intellectuele Biografie* [in Dutch]. Amsterdam: Ambo, 2006.

Root, Margaret Cool. *A Scientist Views the Past: The Samuel A. Goudsmit Collection of Egyptian Antiquities*. Ann Arbor, Michigan: Kelsey Museum of Archaeology, 1982.

Rosbottom, Ronald. *When Paris Went Dark: The City of Light Under German Occupation, 1940–44*. London: John Murray, 2014.

Rossem, Maarten van. *Drie oorlogen* [in Dutch]. Amsterdam: Nieuw Amsterdam, 2007.

Schweber, Silvan. *In the Shadow of the Bomb: Oppenheimer, Bethe, and the Moral Responsibility of the Scientist*. Princeton, NJ: Princeton University Press, 2000.

———. *Einstein & Oppenheimer: The Meaning of Genius*. Cambridge, MA: Harvard University Press, 2008.

BIBLIOGRAPHY

———. *Nuclear Forces. The Making of the Physicist Hans Bethe.* Cambridge, MA: Harvard University Press, 2012.

Smith, P. D. *Doomsday Men: The Real Dr. Strangelove and the Dream of the Superweapon.* London: Allen Lane, 2007.

Speer, Albert. *De Derde Rijk-dagboeken* [in Dutch]. Amsterdam: Meulenhoff, 2015.

Susskind, Leonard. *Quantum Mechanics: The Theoretical Minimum.* New York: Basic Books, 2014.

Talalay, Lauren, and Margaret Cool Root. *Passionate Curiosities: Tales of Collectors and Collections from the Kelsey Museum.* Ann Arbor, Michigan: Kelsey Museum of Archaeology, 2015.

Teller, Edward. *Memoirs: A Twentieth-Century Journey in Science and Politics.* Oxford: Perseus, 2001.

INDEX